RESPONDING TO Oil Spills IN THE U.S. ARCTIC MARINE ENVIRONMENT

Committee on Responding to Oil Spills in the U.S. Arctic Marine Environment

Ocean Studies Board
Division of Earth and Life Studies

Polar Research Board
Division of Earth and Life Studies

Marine Board
Transportation Research Board

NATIONAL RESEARCH COUNCIL
OF THE NATIONAL ACADEMIES

THE NATIONAL ACADEMIES PRESS
Washington, D.C.
www.nap.edu

THE NATIONAL ACADEMIES PRESS • 500 Fifth Street, NW • Washington, DC 20001

NOTICE: The project that is the subject of this report was approved by the Governing Board of the National Research Council, whose members are drawn from the councils of the National Academy of Sciences, the National Academy of Engineering, and the Institute of Medicine. The members of the committee responsible for the report were chosen for their special competences and with regard for appropriate balance.

Funding for this study was provided by the U.S. Arctic Research Commission, the American Petroleum Institute under grant number 2011-105958, the U.S. Coast Guard under cooperative agreement number DTMA1H11001, U.S. Department of the Interior, Bureau of Ocean Energy Management under purchase order number M11PX00116 and Bureau of Safety and Environmental Enforcement under purchase order number E12PX00061, the Marine Mammal Commission under purchase order number DC-260-79EC085782, the National Oceanic and Atmospheric Administration under contract number WC133R-11-CQ-0048, the Oil Spill Recovery Institute under grant number 12-10-02, and the National Academy of Sciences. Any opinions, findings, conclusions, or recommendations expressed in this publication are those of the authors and do not necessarily reflect the views of the organizations or agencies that provided support for the project.

International Standard Book Number-13: 978-0-309-29886-5
International Standard Book Number-10: 0-309-29886-5
Library of Congress Catalog Card Number: 2014942825

Cover photograph provided by Richard Glenn, Arctic Slope Regional Corporation.

Additional copies of this report are available for sale from the National Academies Press, 500 Fifth Street, NW, Keck 360, Washington, DC 20001; (800) 624-6242 or (202) 334-3313; Internet, http://www.nap.edu/.

Printed in the United States of America

THE NATIONAL ACADEMIES

Advisers to the Nation on Science, Engineering, and Medicine

The **National Academy of Sciences** is a private, nonprofit, self-perpetuating society of distinguished scholars engaged in scientific and engineering research, dedicated to the furtherance of science and technology and to their use for the general welfare. Upon the authority of the charter granted to it by the Congress in 1863, the Academy has a mandate that requires it to advise the federal government on scientific and technical matters. Dr. Ralph J. Cicerone is president of the National Academy of Sciences.

The **National Academy of Engineering** was established in 1964, under the charter of the National Academy of Sciences, as a parallel organization of outstanding engineers. It is autonomous in its administration and in the selection of its members, sharing with the National Academy of Sciences the responsibility for advising the federal government. The National Academy of Engineering also sponsors engineering programs aimed at meeting national needs, encourages education and research, and recognizes the superior achievements of engineers. Dr. C. D. Mote, Jr., is president of the National Academy of Engineering.

The **Institute of Medicine** was established in 1970 by the National Academy of Sciences to secure the services of eminent members of appropriate professions in the examination of policy matters pertaining to the health of the public. The Institute acts under the responsibility given to the National Academy of Sciences by its congressional charter to be an adviser to the federal government and, upon its own initiative, to identify issues of medical care, research, and education. Dr. Victor J. Dzau is president of the Institute of Medicine.

The **National Research Council** was organized by the National Academy of Sciences in 1916 to associate the broad community of science and technology with the Academy's purposes of furthering knowledge and advising the federal government. Functioning in accordance with general policies determined by the Academy, the Council has become the principal operating agency of both the National Academy of Sciences and the National Academy of Engineering in providing services to the government, the public, and the scientific and engineering communities. The Council is administered jointly by both Academies and the Institute of Medicine. Dr. Ralph J. Cicerone and Dr. C. D. Mote, Jr., are chair and vice chair, respectively, of the National Research Council.

www.national-academies.org

COMMITTEE ON RESPONDING TO OIL SPILLS IN ARCTIC MARINE ENVIRONMENTS

MARTHA R. GRABOWSKI, *Chair*, Le Moyne College, Syracuse, New York, and Rensselaer Polytechnic Institute, Troy, New York
THOMAS COOLBAUGH, ExxonMobil Research and Engineering, Fairfax, Virginia
DAVID F. DICKINS, DF Dickins Associates, LLC, La Jolla, California
RICHARD GLENN, Arctic Slope Regional Corporation, Barrow, Alaska
KENNETH LEE, Commonwealth Scientific and Industrial Research Organisation, Australia
WILLIAM (LEE) MAJORS, Alaska Clean Seas, Prudhoe Bay
MARK D. MYERS, University of Alaska, Fairbanks
BRENDA L. NORCROSS, University of Alaska, Fairbanks
MARK REED, SINTEF, Norway
BRIAN SALERNO,[1] BIMCO, Washington, D.C.
ROBERT SUYDAM, North Slope Borough, Barrow, Alaska
JAMES M. TIEDJE (NAS), Michigan State University, East Lansing
MARY LOUISE TIMMERMANS, Yale University, New Haven, Connecticut
PETER WADHAMS, Cambridge University, United Kingdom

POLAR RESEARCH BOARD LIAISONS

MOLLY McCAMMON, Alaska Ocean Observing System, Anchorage
CARYN REA, ConocoPhillips, Anchorage, Alaska

NATIONAL RESEARCH COUNCIL STAFF

DEBORAH GLICKSON, Senior Program Officer
LAUREN BROWN, Associate Program Officer, Polar Research Board
STACEE KARRAS, Research Associate
HEATHER CHIARELLO, Senior Program Assistant (until April 2013)
PAYTON KULINA, Program Assistant (from June 2013)

[1] Resigned from the committee.

Preface

Balance. This is an important word in the Arctic, an area that serves as an integrator of many of the Earth's large-scale systems and processes, and also an area where choices made have substantial impact on the Arctic and its neighbors. Many competing forces coexist and collide in the Arctic: harsh environmental conditions, economic drivers, science and technology capabilities, logistical and infrastructure challenges, ecosystem protection needs, food security concerns, and the needs of traditional cultures and societies. Balancing the needs and requirements of these forces is part of the challenge and opportunity presented in the complex, large-scale system that is the Arctic.

Within this context, the National Research Council was asked by eight sponsors who represent many of these drivers to consider the adequacy and sufficiency of resources, technology, research, human resources, funding, and logistics to respond to an Arctic oil spill. The committee sought to balance in its work traditional and scientific knowledge of the Arctic and of oil spill response operations, engineering, technology, policies, procedures, and equipment. It considered the needs and concerns of the committee's sponsors; government, public, private, for-profit and not-for-profit organizations; citizens and organizations with Arctic interests; and the needs and interests of Arctic inhabitants. The committee also considered lessons learned from events and case studies from oil spill response efforts around the world.

The committee's work was enhanced by the participation and input provided by a number of individuals, organizations, and groups, many of whom are listed elsewhere in this report. The committee solicited input from workshop participants, speakers, and experts across the spectrum of traditional knowledge, science, engineering, vessel and oil spill operations, and regulatory and government affairs. The committee's work was also enhanced by the insight, experience, and collegiality of its globally distributed members, as it followed the tenets of an earlier National Research Council (1996) report, to "get the science right and get the right science; to get the participation right and get the right participation; and to develop an inclusive and thoughtful analytic-deliberative process." The result is a report that considers the adequacy of and needs for oil spill response in the U.S. Arctic, drawing on the wisdom and expertise of many in and of the Arctic, and that considers significant challenges in an important ecosystem.

It was my privilege to work with our committee; our project sponsors; our study director,

Deb Glickson; Polar Research Board Associate Program Officer Lauren Brown; Ocean Studies Board Director Susan Roberts; Marine Board Directors Joedy Cambridge and Scott Brotemarkle; and the rest of the National Academies staff during the course of this study. Thank you all for sharing your wisdom and insight. May we meet again in future endeavors.

M. Grabowski, *Chair*
Committee on Responding to Oil Spills in the
U.S. Arctic Marine Environment

Acknowledgments

This report was greatly enhanced by the participants of meetings held as part of this study. The committee would like to acknowledge those who gave presentations at committee meetings: Bill Adams (Remote Energy Security Technologies Collaborative [RESTCo]), Doug Baird (National Oceanic and Atmospheric Administration [NOAA]), Geoff Baker (Crowley Maritime Corporation), Mary Baker (NOAA), Lawson Brigham (University of Alaska, Fairbanks [UAF]), Christy Bohl (Bureau of Safety and Environmental Enforcement), Gene Brooks (Maersk Line, Ltd.), Harry Brower, Jr. (North Slope Borough Department of Wildlife Management), Larry Dietrick (Alaska Department of Environmental Conservation), Hajo Eicken (UAF), Michael Faust (ConocoPhillips), Jeffrey Ferguson (NOAA), Adrian Gall (ABR, Inc.), Larry Hinzman (UAF), Charles Hopson, John Hopson, Jr. (Wainwright Public Works), Christopher Ives (RESTCo), Christopher Krenz (Oceana), Nettie La Belle-Hamer (UAF), Joe LoSciuto (ASRC Energy Services), Joe Mello Leavitt, Amy Merten (NOAA), Vince Mitchell (Lamor Corporation), RADM Thomas Ostebo (U.S. Coast Guard), Ed Owens (Owens Coastal Consultants, Ltd.), Ed Page (Marine Exchange of Alaska), Shirish Patil (UAF), Vladimir Romanovsky (UAF), Stan Senner (Ocean Conservancy), Gay Sheffield (UAF), Kirk Sherwood (Bureau of Ocean Energy Management), Brad Smith (NOAA), Mark Swanson (Prince William Sound Regional Citizens' Advisory Council), Fran Ulmer (U.S. Arctic Research Commission), Peter van Tuyn (World Wildlife Fund), Peter Velez (Peter Velez Engineering LLC), Glen Watabayashi (NOAA), Thomas Weingartner (UAF), and Peter Winsor (UAF).

The committee would also like to thank Karissa Goessl and Patrick Curtin of LeMoyne College, who assisted at the committee's third meeting in Fairbanks, Alaska.

This report has been reviewed in draft form by individuals chosen for their diverse perspectives and technical expertise, in accordance with procedures approved by the National Research Council's Report Review Committee. The purpose of this independent review is to provide candid and critical comments that will assist the institution in making its published report as sound as possible and to ensure that the report meets institutional standards for objectivity, evidence, and responsiveness to the study charge. The review comments and draft manuscript remain confidential to protect the integrity of the deliberative process. We wish to thank the following individuals for their participation in their review of this report:

PER JOHAN BRANDVIK, SINTEF Marine Environmental Technology
LAWSON BRIGHAM, University of Alaska, Fairbanks
BILL EICHBAUM, World Wildlife Fund
JOHN FARRINGTON, Woods Hole Oceanographic Institution
JACQUELINE GREBMEIER, University of Maryland
MOLLY McCAMMON, Alaska Ocean Observing System
HUMPHREY MELLING, Fisheries and Oceans Canada
JOSEPH MULLIN, Joseph Mullin Consulting
PARTHA PATRA, Columbia University
STEPHEN POTTER, SL Ross Environmental Research Ltd.
PONISSERIL SOMASUNDARAN (NAE), Columbia University
WILFORD WEEKS (NAE), University of Alaska, Fairbanks (emeritus)

Although the reviewers listed above have provided many constructive comments and suggestions, they were not asked to endorse the conclusions or recommendations nor did they see the final draft of the report before its release. The review of this report was overseen by **RADM Malcolm MacKinnon (NAE)**, MacKinnon-Searle Consortium LLC, appointed by the Divison on Earth and Life Studies, and **Bonnie McCay (NAS)**, Rutgers University, appointed by the Report Review Committee, who were responsible for making certain that an independent examination of this report was carried out in accordance with institutional procedures and that all review comments were carefully considered. Responsibility for the final content of this report rests entirely with the authoring committee and the institution.

Contents

Scenarios, Boxes, Figures, and Tables

SCENARIOS

BOXES

FIGURES

TABLES

Summary

The Arctic system serves as an integrator for the Earth's physical, biological, oceanic, and atmospheric processes, with impacts beyond the Arctic itself. The risk of an oil spill in the Arctic presents hazards for Arctic nations and their neighbors.

The threat of a major Arctic oil spill and the potential impacts on the region's marine ecosystems are of concern for a broad range of U.S. and international interests, including Alaskan natives and others who live in the region, citizens and organizations concerned about the health of the Arctic environment, agencies committed to protecting the environment and threatened species, agencies that regulate extractive activities or transportation, and industries that plan to develop oil and gas, shipping routes, fisheries, or tourism.

Rapid climate change is leading to retreat and thinning of Arctic sea ice, potentially increasing the accessibility of U.S. Arctic marine waters for commercial activities. With this projected rise in activity come additional concerns about the risk of oil spills. Recent interest in developing the rich oil and gas resources in federal waters offshore of Alaska has led to planning, environmental assessments, and preliminary drilling for oil and gas exploration. In addition, expanding maritime activity in the region includes the potential for greater seasonal use by tankers and bulk carriers, fishing fleets that follow the northward migration of fish stocks, and cruise ships interested in exploiting the public's desire to interact with Arctic wilderness.

The National Research Council was asked by the American Petroleum Institute, the U.S. Arctic Research Commission, the Bureau of Safety and Environmental Enforcement (BSEE), the Bureau of Ocean Energy Management (BOEM), the U.S. Coast Guard (USCG), the Marine Mammal Commission, the National Oceanic and Atmospheric Administration (NOAA), and the Oil Spill Recovery Institute to assess the current state of science and engineering regarding oil spill response and Arctic marine environments, with emphasis on potential impacts in U.S. waters in the Bering Strait and Chukchi and Beaufort Seas. The committee was tasked to review research activities and recommend strategies to advance research and address information gaps, to identify opportunities and constraints for advancing oil spill research, to describe promising new concepts and technologies, and to assess the types of baseline information needed to monitor the impacts of an oil spill and to develop plans for recovery and restoration.

THE ARCTIC ENVIRONMENT

Arctic oil spill response is challenging because of extreme weather and environmental conditions; the lack of existing or sustained communications, logistical, and information infrastructure; significant geographic distances; and vulnerability of Arctic species, ecosystems, and cultures.

A fundamental understanding of the dynamic Arctic region (Figures S.1 and S.2) is needed to help guide oil spill response and recovery efforts. Information on physical processes—including ocean

Figure S.1 Location map of Alaska and the continental United States, and surrounding countries and water bodies. The red box shows the location of the inset map in Figure S.2. Bathymetry, geopolitical boundaries, capitals, and select Alaskan cities are also shown.

Figure S.2 Location map of Alaska and U.S. Arctic waters, focused on the Bering Strait, Chukchi Sea, and Beaufort Sea. Geopolitical boundaries, principal coastal communities, cities, and bathymetry are also shown. Map area corresponds to the red box in Figure S.1.

circulation, ice cover, marine weather, and coastal processes—is important to frame the environmental context for the Arctic ecosystem and can help responders predict where oil will spread and how weathering might change its properties. Parameters such as air and water temperature, wind velocity, and hours of daylight are important considerations in choosing an effective and safe response strategy.

Knowledge of ice thickness, concentration, and extent is essential for anticipating the likely behavior of oil in, under, and on ice and determining applicable response strategies, while high-quality bathymetry, nautical charting, and shoreline mapping data are needed for marine traffic management and oil spill response. From a biological perspective, understanding population dynamics and interconnections within the Arctic food web will enable the determination of key species that are most important for monitoring in the instance of an oil spill.

Baseline data are critical to assess changes over time. In the Arctic, historical data do not provide reliable baselines to assess current environmental or ecosystem states, nor can they fully anticipate potential impacts due to factors such as seasonal and interannual variations or climate change. Instead, monitoring approaches will need to take advantage of benchmarks, or reference points over time, rather than static baselines. Critical types of benchmark data for oil spill response in the Arctic include:

- Spatial and temporal distributions and abundances for fishes, birds, and marine mammals;
- Subsistence and cultural use of living marine resources;
- Identification and monitoring of areas of biological significance;
- Rates of change for key species;
- Sensitivity of key Arctic species to hydrocarbons;
- High-resolution coastal topography and shelf bathymetry; and
- Measurements of ice cover, thickness, and distribution.

Additional research and development needs include meteorological-ocean-ice forecast model systems at high temporal and spatial resolutions and better assimilation of traditional knowledge of sea state and ice behavior into forecasting models. Releasing proprietary monitoring data from exploration activities would increase knowledge of Arctic benchmark conditions. When appropriate, Arctic communities could also release data that they hold regarding important sites for fishing, hunting, and cultural activities.

In many instances, frequent and regular long-term monitoring will be needed to determine trends. Because data are or will be collected by a number of local, state, and federal agencies, as well as industry and academia, a complete information system that integrates Arctic data in support of oil spill preparedness, response, and restoration and rehabilitation is needed. Achieving this goal requires the development of international standards for Arctic data collection, sharing, and integration. A long-term, community-based, multiuse Arctic observing system could provide critical data at a variety of scales.

Recommendation: A real-time Arctic oceanographic-ice-meteorological forecasting system is needed to account for variations in sea ice coverage and thickness and should include patterns of ice movement, ice type, sea state, ocean stratification and circulation, storm surge, and improved resolution in areas of potential risk. Such a system requires robust, sustainable, and effective acquisition of relevant observational data.

Recommendation: High-resolution satellite and airborne imagery needs to be coupled with up-to-date high-resolution digital elevation models and updated regularly to capture the dynamic, rapidly changing U.S. Arctic coastline. Nearshore bathymetry and topography should be collected at a scale appropriate for accurate modeling of coastline vulnerability and storm surge sensitivity. Short- and long-term Arctic nautical charting and shoreline mapping that have been identified in NOAA and U.S. Geological Survey plans should be adequately resourced, so that mapping efforts can be initiated, continued, and completed in timescales

relevant to anticipated changes. To be effective, Arctic mapping priorities should continue to be developed in consultation with stakeholders and industry and should be implemented systematically rather than through surveys of opportunity.

OIL SPILL RESPONSE RESEARCH

A comprehensive, collaborative, long-term Arctic oil spill research and development program that integrates all knowledgeable sectors and focuses on oil behavior, response technologies, and controlled field releases is needed.

Laboratory experiments, field research, and practical experience gained from responding to past oil spills have built a strong body of knowledge on oil properties and oil spill response techniques. However, much of the work has been done for temperate regions, and there are areas where additional research is needed to make informed decisions about the most effective response strategies for different Arctic situations. In the presence of lower water temperatures or sea ice, the processes that control oil weathering—such as spreading, evaporation, photo-oxidation, emulsification, and natural dispersion—are slowed down or eliminated for extended periods of time. Because of encapsulation of oil by new ice growth, oil can also be separated from the environment for months at a time. Understanding how oil behaves or changes in the Arctic environment can help define the most effective oil spill response actions.

In addition to ongoing research on oil properties and weathering in high latitudes, there is a need to validate current and emerging oil spill response technologies on operational scales under realistic environmental conditions. Carefully planned and controlled field releases of oil in the U.S. Arctic would improve the understanding of oil behavior in the Bering Strait and Beaufort and Chukchi Seas and allow for the evaluation of new response strategies specific to the region. Scientific field releases that have been conducted elsewhere in the Arctic demonstrate that such studies can be carried out without measureable harm to the environment.

Recommendation: A comprehensive, collaborative, long-term Arctic oil spill research and development program needs to be established. The program should focus on understanding oil spill behavior in the Arctic marine environment, including the relationship between oil and sea ice formation, transport, and fate. It should include assessment of oil spill response technologies and logistics, improvements to forecasting models and associated data needs, and controlled field releases under realistic conditions for research purposes. Industry, academic, government, non-governmental, grassroots, and international efforts should be integrated into the program, with a focus on peer review and transparency. An interagency permit approval process that will enable researchers to plan and execute deliberate releases in U.S. waters is also needed.

Oil Spill Countermeasures

Key response countermeasures and tools for oil removal in Arctic conditions include biodegradation (including oil treated with dispersants), in situ burning, chemical herders, mechanical containment and recovery, detection and tracking, and oil spill trajectory modeling. These are joined by the "no response" option of natural recovery, which is a viable alternative in some situations. No single technique will apply in all situations. The oil spill response toolbox requires flexibility to evaluate and apply multiple response options, if necessary. Well-defined and well-tested decision processes are critical to expedite review and approval of countermeasure options in emergency situations.

Biodegradation and Dispersants

Biodegradation of petroleum hydrocarbons by naturally occurring microbial communities is a major process contributing to the eventual removal of oil that enters the marine environment. Recent studies suggest that indigenous bacteria in Arctic waters degrade oil faster than previously thought and that biodegradation is not strongly inhibited by cold water temperatures. Current research is focused on better understanding of this environmentally important process.

Chemical dispersants facilitate the dilution of oil in the water column and promote biodegradation. There has been considerable debate over the effectiveness of chemical dispersants at low seawater temperatures, but recent studies have shown that dispersants can be effective on nonemulsified oil at freezing temperatures if viscosity does not increase significantly.

Subsea injection of dispersant is a promising option for mitigation of oil spills from a wellhead blowout and could disperse oil at higher rates and with higher efficiency than aerial application. Subsea injection can also operate independently of darkness, extreme temperatures, strong winds, rough seas, or the presence of ice. However, more work needs to be done on the effectiveness, systems design, and short- and long-term impacts of subsea dispersant delivery.

Recommendation: Dispersant pre-authorization in Alaska should be based on sound science, including research on fates and effects of chemically dispersed oil in the Arctic environment, experiments using oils that are representative of those in the Arctic, toxicity tests of chemically dispersed oil at realistic concentrations and exposures, and the use of representative microbial and lower-trophic benthic and pelagic Arctic species at appropriate temperatures and salinities.

In Situ Burning

In situ burning is a viable spill response countermeasure in the Arctic. Ice can often provide a natural barrier to maintain the necessary oil thicknesses for ignition, without the need for booms. With relatively fresh oil that is wind herded against an ice edge, or collects in melt pools in the spring, removal efficiencies in excess of 90% are achievable through in situ burning. However, in very open drift ice conditions, oil spills can rapidly spread too thinly to ignite. To improve the limits of in situ burning, further research is needed to evaluate improved ignition methods and to explore the use of aerially applied oil-herding chemicals at different spatial scales and with different oil types, including weathered states.

Mechanical Containment and Recovery

Mechanical containment and recovery removes oil from the marine environment, rather than adding chemicals or generating burn residue. However, when dealing with large offshore spills, the oil can quickly spread to a thin sheen, which makes it difficult to achieve a significant rate of recovery. Large quantities of containment boom and hundreds of vessels and skimmers are needed to concentrate thin, rapidly spreading oil slicks. The lack of approved disposal sites on land for contaminated water and waste, lack of port facilities to accept deep-draft vessels, and limited airlift capability to remote communities complicates the large-scale use of mechanical containment and recovery to respond to Arctic spills. Mechanical recovery can provide a viable option for small, contained spills in pack ice, or for larger spills under fast ice. Arctic-relevant mechanical recovery improvements include cold temperature operability and independent propulsion; however, response to a large offshore spill in the U.S. Arctic is unlikely to rely only upon mechanical containment and recovery because of its inefficiency.

Detection, Monitoring, and Modeling

To mount an effective response, it is critical to know where spilled oil is at any given time. Over the past decade, several large government and industry programs have evaluated the variety of rapidly developing remote sensing technologies used for detection, including sonar, synthetic aperture radar, infrared, and ground-penetrating radar. In addition, the use of unmanned aerial vehicles and autonomous underwater vehicles for oil detection and tracking has grown. However, there will always be a need for aerial observers to map oiled areas and transmit critical information to response crews. Detection methods work hand-in-hand with advanced oil spill trajectory modeling to understand where oil is moving. Promising advances in modeling have accounted for the incorporation of oil into brine channels as well as the bulk freezing of oil into ice, although better modeling of under-ice roughness is still needed. Investment in detection and response strategies for oil on, within, and trapped under ice will be necessary for contingency planning. In addition, robust operational meteorological-ocean-ice and oil spill trajectory forecasting models for the U.S. Arctic would further improve oil spill response efforts.

Arctic oil spill research and development needs for improved decision support include:

- Improving methods for in situ burning, dispersant application, and use of chemical herders;
- Understanding limitations of mechanical recovery in both open water and ice;
- Investing in under-ice oil detection and response strategies;
- Integrating remote sensing and observational techniques for detecting and tracking ice and oil;

- Determining and verifying biodegradation rates for hydrocarbons in offshore environments;
- Evaluating the toxicity of dispersants and chemically dispersed oil on key Arctic marine species; and
- Summarizing relevant ongoing and planned research worldwide to achieve synergy and avoid unnecessary duplication.

Recommendation: Priorities for oil spill research should leverage existing joint agreements and be addressed through a comprehensive, coordinated effort that links industry, government, academia, international and local experts, and non-governmental organizations. The Interagency Coordinating Committee on Oil Pollution Research, which is tasked to coordinate oil spill research and development among agencies and other partners, should lead the effort.

OPERATIONS, LOGISTICS, AND COORDINATION FOR AN ARCTIC OIL SPILL

Marine activities in U.S. Arctic waters are increasing without a commensurate increase in the logistics and infrastructure needed to conduct these activities safely. As oil and gas, shipping, and tourism activities increase, the U.S. Coast Guard will need an enhanced presence and performance capacity in the Arctic. U.S. support for Arctic missions, including oil spill response, requires significant investment in infrastructure and capabilities.

U.S. Coast Guard Needs

The USCG has a low level of presence in the Arctic, especially during the winter. USCG personnel, equipment, transportation, communication, navigation, and safety resources needed for oil spill response are not adequate for overseeing oil spill response in the Arctic, and the Coast Guard's efforts to support Arctic oil spill planning and response in the absence of a dedicated and adequate budget are admirable but inadequate.

Recommendation: As oil and gas, shipping, and tourism activities increase, the USCG will need an enhanced presence and performance capacity in the Arctic, including area-specific training, icebreaking capability, improved availability of vessels for responding to oil spills or other emergency situations, and aircraft and helicopter support facilities for the open water season and eventually year round. Furthermore, Arctic assignments for trained and experienced personnel and tribal liaisons should be of longer duration, to take full advantage of their skills. Sustained funding will be needed to increase the USCG presence in the Arctic and to strengthen and expand its ongoing Arctic oil spill research programs.

Vessel traffic is not actively managed in the Bering Strait or in the U.S. Arctic, nor is there a comprehensive system for real-time traffic monitoring. The lack of a U.S. vessel traffic monitoring system for the Arctic creates significant vulnerability for missions including oil spill response and creates undue reliance on private industry and foreign national systems. Private receivers are used to track vessels in the Bering Strait and along a large part of Alaskan coastal areas, but there are significant gaps in coverage. Consequently, there are numerous regional "blind spots" where an early indication of elevated risks may not be apparent to officials ashore.

Recommendation: The USCG should expedite its evaluation of traffic through the Bering Strait to determine if vessel traffic monitoring systems, including an internationally recognized traffic separation scheme, are warranted. If so, this should be coordinated with Russia. The USCG should also consider obtaining broader satellite monitoring of Automatic Identification System signals in the Arctic through government means or from private providers.

Infrastructure

The lack of infrastructure in the Arctic would be a significant liability in the event of a large oil spill. Communities are dependent on air and seasonal marine transport for the movement of people, goods, and services, and there are few equipment caches with boom, dispersants, and in situ burn materials available for the North Slope and Northwest Arctic Boroughs. It is unlikely that responders could quickly react to an oil spill unless there were improved port and air access, stronger supply chains, and increased capacity to handle equipment, supplies, and personnel. Prepositioning a suite of response equipment throughout the Arctic, including aerial in situ burn and dispersant capability, would provide immediate access to a number of rapid response oil spill countermeasure options.

Building U.S. capabilities to support oil spill response will require significant investment in physical infrastructure and human capabilities, from communications and personnel to transportation systems and traffic monitoring. Human and organizational infrastructure improvements are also required to improve international and tribal partnerships so as to leverage scientific and traditional knowledge and best practices. A truly capable end-to-end system for oil spill response would require integration of Arctic data in support of preparedness, response, and restoration and rehabilitation.

There is presently no funding mechanism to provide for development, deployment, and maintenance of temporary and permanent infrastructure. One approach to provide a funding mechanism for infrastructure development and oil spill response operations would be to enable a public-private-municipal partnership to receive a percentage of lease sale revenues, rents, bonuses, or royalty payments that are currently deposited in the federal treasury.

Recommendation: Infrastructure to support oil spill response should be enhanced in the North Slope Borough, Northwest Arctic Borough, and communities along the Bering Strait,[1] with marine facilities for addressing response operations. The scope, scale, and location of infrastructure needs should be determined through structured decision processes, studies, and risk assessments.

[1] The wording of this recommendation was edited after release of the prepublication to explicitly include communities along the Bering Strait.

For spills occurring within U.S. jurisdiction, the Oil Pollution Act of 1990 provides the necessary legal framework for the responsible party—often the owners or operators of energy or shipping companies—to fund response operations and provide compensation for damages. The burden of cost can fall on the government, however, when the cost of oil spill response exceeds liability limits or when the responsible party cannot be found. The Oil Spill Liability Trust Fund, a fund maintained by the federal government for these situations, may prove insufficient to cover the sociological and economic damages of affected communities. A "whole government" approach that includes the ability to deal with broad societal impacts of a spill may be necessary.

Training and Organization

Local communities possess in-depth knowledge of ice conditions, ocean currents, and marine life in areas that could be affected by oil spills. Failure to include local knowledge during planning and response may increase the risk of missing significant environmental information, yet there appears to have been only modest efforts to integrate local knowledge into formal incident command-based responses. Developing and maintaining trained village response teams integrates local knowledge and utilizes existing human resources for effective oil spill response. The North Slope Borough has a well-developed local emergency response team, and the Northwest Arctic Borough is strengthening this capability in its region.

Recommendation: The USCG and Alaska Department of Environmental Conservation should undertake the development of an oil spill training program for local entities so as to develop trained response teams in local villages. Industry should continue to participate in local training initiatives. Local officials and trained village response teams should be included in the coordinated decision-making and command process during a response event. Input from community experts should be actively solicited for inclusion in response planning and considered in conjunction with data derived from other sources. The USCG should set this as an exercise objective in all government-led oil spill response exercises in the Arctic and should set the expectation that industry-led exercises will do the same.

Flexible and scalable organization is important to develop an effective Arctic oil spill response. This can be achieved through drills, case studies, simulation, and organizational learning. To build the systemwide capacity to respond to large-scale, distributed Arctic oil spill response, sustained long-term training and continued resource investments are required. Inclusive and trustful communications, relationship building, and decision making, clear accountability, and ongoing assessment and improvement are also necessary.

Recommendation: Relevant federal, state, and municipal organizations (such as USCG, NOAA, BSEE, BOEM, Alaska Department of Environmental Conservation, Alaska Department of Natural Resources, U.S. Fish and Wildlife Service, Alaska Department of Fish and Game, North Slope Borough, and Northwest Arctic Borough), local experts,

industry, and academia should undertake regularly scheduled oil spill exercises designed to test and evaluate the flexible and scalable organizational structures needed for highly reliable Arctic oil spill response.

International Coordination

The United States has long engaged its regional neighbors in Arctic spill preparedness and has bilateral agreements with both Canada and Russia regarding oil spill response. Formal contingency planning and exercises with Canada have enabled both the United States and Canada to refine procedures and legal requirements for cross-border movement of technical experts and equipment in the event of an emergency. Continuing to exercise the bilateral agreement with Russia will enable both countries to address practical issues that could arise in an actual spill. An active exercise program with Russia, similar to that with Canada, could identify problems and resolve them in advance.

The Russian Federation's commitment to the economic development of the Northern Sea Route has expanded the volume of large-vessel traffic through the Bering Strait, suggesting increased risks of a major vessel accident and implications for environmental impacts in U.S. waters. The resolution of anticipated response problems, including communications between command centers, coordinated planning, transboundary movement of people and equipment, and identification of translators, needs to be accomplished in advance of an actual event.

Recommendation: The USCG should expand its bilateral agreement with Russia to include Arctic spill scenarios and conduct regularly scheduled exercises to establish joint responses under Arctic conditions and should build on existing bilateral agreements with Russia and Canada to develop and exercise a joint contingency plan.

STRATEGIES FOR RESPONSE AND MITIGATION

Oil spill response effectiveness could be improved by adopting decision processes such as Net Environmental Benefit Analysis, by developing inclusive organizational response practices in advance of an event, and by enhancing resource availability for training, infrastructure, and monitoring.

All pre-spill strategies emphasize oil spill prevention above everything else. In the event of an oil spill, however, strategies for decision making and response are critical in order to keep oil away from the shore and to minimize impacts on sensitive habitats, organisms, and people. There are no response methods that are completely effective or risk free. Decision processes that evaluate options and response strategies are critical to an effective response. The Net Environmental Benefit Analysis (NEBA) process provides a framework to determine which oil spill countermeasures will be the most effective and will cause the least ecological damage, based on an analysis of environmental tradeoffs. NEBA incorporates prioritization criteria for the protection of sensitive and important ecosystem components that could be impacted by oiling, cleanup operations, or residual oil—for example, marine mammals, coastal habitats, fishes, or areas of cultural significance. An Arctic NEBA would also include information on the transport, fate, and potential effects of the spilled oil; knowledge of operational limits, advantages, and disadvantages of each oil spill response countermeasure for the natural resources at risk; and consideration of logistical constraints and cleanup intensity. Because of the range of conditions typically encountered within an area affected by an oil spill, it is likely that a combination of countermeasures, rather than a single response option, would be most likely to provide optimal protection for all environmental resources.

Recommendation: A decision process such as NEBA should be used to select the response options that offer the greatest overall reduction of adverse environmental impacts. In the Arctic, areas of cultural and subsistence importance should be among the priority ecosystem components. In light of concerns regarding detrimental effects on ecosystems, further study should focus on the impact of oil spills on Arctic food webs and dynamics at different trophic levels. The process should involve regulators, resource managers, health authorities, technical specialists, scientific experts, and local experts.

The potential impact of oil and countermeasures on wildlife is a major concern during an oil spill response. Controlling oil release and spread at the source of a spill, deterring animals from entering oiled areas, and capturing and rehabilitating oiled wildlife can help minimize the potential impact on wildlife, the broader ecosystem, and the food web. However, rehabilitation and release in the Arctic are complicated by remote locations, lack of response equipment, concerns over subsistence use of potentially oiled animals, and safety considerations when dealing with large animals such as polar bears and walruses. There is a general lack of scientific study, approved protocols, and consensus among decision makers regarding marine mammal deterrence. Wildlife response plans will need to include key indicators of environmental health, and prioritize response strategies. This includes a no-response strategy, which may be preferable for some species.

Recommendation: The U.S. Fish and Wildlife Service, NOAA's National Marine Fisheries Service, the Alaska Department of Fish and Game, co-management organizations, and local government and communities are the trustees for wildlife deterrence and rehabilitation. As appropriate, these agencies and groups should work together with industry to explore and improve deterrence and rehabilitation methods for wildlife. Additional research and development for improved methods could benefit from the involvement of universities, non-governmental organizations, and others. Priorities should be set and regularly updated by the trustees for oil spill response based on the type of wildlife threatened, the season, other factors related to a spill, and updated research and methodology.

1 Introduction

A CHANGING ARCTIC

The Arctic acts as an integrating, regulating, and mediating component of the physical, atmospheric, and cryospheric systems that govern life on Earth. It is also undergoing rapid climate change, the rate of which is projected to accelerate in coming decades (e.g., Serreze et al., 2000; Bitz et al., 2012; Kay et al., 2012; Jeffries and Richter-Menge, 2013). Surface air temperature increases in the Arctic in recent decades are about two to four times larger than observed in the midlatitudes, with evidence that the increase will continue (Overland et al., 2012, 2013). This "Arctic amplification" has been attributed to various complicated interactions between physical mechanisms (NRC, 2014), including albedo (solar reflectance) change due to sea ice and snowline retreat and latitudinal differences in surface energy radiation (e.g., IPCC, 2007, 2013; Francis and Vavrus, 2012; Pithan and Mauritsen, 2014).

The most obvious evidence for Arctic change has been the well-documented retreat and thinning of Arctic sea ice cover (e.g., Stroeve et al., 2012; Wadhams, 2012a,b; Overland and Wang, 2013; Perovich et al., 2013; IPCC, 2014). Between 1979 and 2013, the linear rate of decline of September ice extent was 13.7% per decade. The largest sea ice losses were documented in the Beaufort and Chukchi regions, particularly in the extremely low summer sea ice years of 2007 and 2012.[1] Recent increases in surface ocean temperatures in the Arctic, particularly in the Beaufort and Chukchi Seas, are related to this sea ice retreat (e.g., Perovich et al., 2008, 2011). Mean surface ocean temperatures in the southern Beaufort Sea in August 2007 and 2012 were more than 2°C warmer than the August mean between 1982 and 2006 (Timmermans et al., 2013).

Warming upper ocean temperatures may lead to increased thawing of offshore and coastal permafrost (Shakhova et al., 2010; Portnov et al., 2013; Whiteman et al., 2013) and coastal erosion, which is exacerbated by sea ice loss and increased sea state, as well as sea-level rise (e.g., Jones et al., 2009b; Barnhart et al., 2011). Increased waves are already a feature of the Alaskan coastal zone (e.g., Barnett et al., 2012). Other known climate change manifestations in the Arctic include changing

[1] See http://www.nsidc.org.

atmospheric circulation patterns and increased cloud cover, related in part to the reduced sea ice extent (e.g., Overland et al., 2013).

The feedbacks and transitions occurring in this region will have significant implications for biodiversity, human benefits from the ecosystem, and other important processes within the Arctic and global system (Nilsson et al., 2013). As the Arctic changes, larger areas are becoming more accessible for shipping, exploration, and resource development, which come with increased concerns for oil spills and other types of potentially harmful incidents that could impact U.S. waters. There have been a number of recent efforts that highlight the importance of the Arctic to national interests (e.g., Holland-Bartels and Pierce, 2011; NSTC, 2013; USARC, 2013; USCG, 2013b).

CHARACTERISTICS OF THE U.S. ARCTIC

The Arctic Research and Policy Act of 1984 (P.L. 98-373) defines the Arctic as "all United States and foreign territory north of the Arctic Circle and all United States territory north and west of the boundary formed by the Porcupine, Yukon, and Kuskokwim Rivers [in Alaska]; all contiguous seas, including the Arctic Ocean and the Beaufort, Bering, and Chukchi Seas; and the Aleutian chain." U.S. Arctic waters north of the Bering Strait and west of the Canadian border encompass a vast area divided into the Chukchi and Beaufort Seas (Figures 1.1 and 1.2). In general, the Chukchi Sea is underlain by a very broad and shallow continental shelf, whereas the Beaufort Sea has a relatively narrow shelf that drops into the deep Canada Basin. These seas are ice covered for much of the year but are increasingly experiencing longer periods and larger areas of open water, with the maximum area of open water occurring in mid-September (Jeffries and Richter-Menge, 2013). Much of this region is located north of the Arctic Circle, experiencing winter months with little to no daylight and summer months with up to 24 hours of daylight.

The Alaskan coastline north of the Bering Strait is sparsely inhabited. According to 2010 census data, the total population of the North Slope Borough was less than 8,000 persons.[2] The largest coastal communities in the Northwest Arctic and North Slope Boroughs are Kotzebue and Barrow, which respectively have more than 3,000 and 4,000 residents. All other villages in the boroughs have populations of fewer than 1,000 residents. The indigenous culture is primarily Iñupiat along the Chukchi and Beaufort Seas. The coastal and some inland communities rely extensively on marine subsistence resources, together with revenues from onshore oil development in the Prudhoe Bay area and commercial chum salmon fishing in Kotzebue Sound.

These seas are home to ecosystems with a wide diversity of marine life. Many marine mammals and seabirds migrate seasonally to the Chukchi and Beaufort areas, with some permanent resident populations of polar bears and seals. Owing to the rapid warming of the Arctic and the associated decrease in sea ice, significant changes are occurring in the habitat, range, and behavior of the marine species that inhabit these waters (CAFF, 2013).

The communities of the Beaufort and Chukchi Seas have limited infrastructure and no deep-water ports. None of the communities have permanent road infrastructure connected to the main

[2] See http://www.north-slope.org/departments/mayorsoffice/census_data_2010.php; http://quickfacts.census.gov/qfd/states/02/02188.html; http://quickfacts.census.gov/qfd/states/02/02185.html.

Figure 1.1 Location map of Alaska and the continental United States, and surrounding countries and water bodies. The red box shows the location of inset maps that are referenced throughout the report (beginning with Figure 1.2). Bathymetry, geopolitical boundaries, capitals, and select Alaskan cities are also shown.

highway systems or large communities in Alaska, although some communities are seasonally connected by ice roads. Instead, the communities are largely dependent on air and seasonal marine transport for the movement of people, goods, and services outside their regions. All coastal communities receive barge shipments during the summer and early autumn open water months. Industrial activities of the region include commercial fishing in Kotzebue, Port Clarence, and Norton Sound;[3] the Red Dog lead and zinc mine north of Kotzebue; and oil and gas fields on the North Slope.

[3] See http://www.adfg.alaska.gov/index.cfm?adfg=fishingCommercial.main.

Figure 1.2 Location map of Alaska and U.S. Arctic waters, focused on the Bering Strait, Chukchi Sea, and Beaufort Sea. Geopolitical boundaries, principal coastal communities, cities, and bathymetry are also shown. Map area corresponds to the red box in Figure 1.1.

INDUSTRIAL ACTIVITIES: OIL, GAS, AND SHIPPING

The global ocean has long been exposed to a significant quantity of petroleum hydrocarbons from natural seeps of oil and gas, marine oil transportation accidents, and operational discharges (NRC, 2003; Yakimov et al., 2007). The Arctic marine environment is rich in oil and gas. There are an estimated 30 billion barrels of technically recoverable undiscovered oil[4] in the U.S. Arctic, which equals approximately one-third of the total resource found in the entire circum-Arctic region (Bird

[4] "Technically recoverable" undiscovered oil refers to reserves that could be produced using current technology, without consideration of economic viability. "Economically recoverable" refers to the amount of technically recoverable oil whose price can cover costs of production (BOEM, 2011; also see http://pubs.usgs.gov/of/2011/1103/).

et al., 2008). The subsurface Chukchi Sea Outer Continental Shelf (OCS) is estimated to contain 11 billion barrels of undiscovered economically recoverable oil and 38 trillion cubic feet of natural gas, while the subsurface Beaufort Sea OCS is estimated to contain 6 billion barrels of undiscovered economically recoverable oil and 11 trillion cubic feet of natural gas (at $110/barrel and $7.83/thousand cubic feet of gas; BOEM, 2011). It is estimated that 80 to 90% of petroleum hydrocarbon entering the Arctic marine environment is from natural seeps (AMAP, 2008). Becker and Manen (1988) reported the presence of oil seeps in the coastal regions of Alaska and estimated submarine seepage to be approximately 1,000 tons/yr. As crude oil seepage has been estimated to be 600,000 metric tons/yr globally (Kvenvolden and Cooper, 2003; NRC, 2003), natural seeps may be among the most important sources of oil entering the ocean.

In the Chukchi Sea, most of the exploration activity occurs relatively far offshore (greater than 80-120 km), roughly equidistant from the villages of Point Lay and Wainwright (Figure 1.3). For this

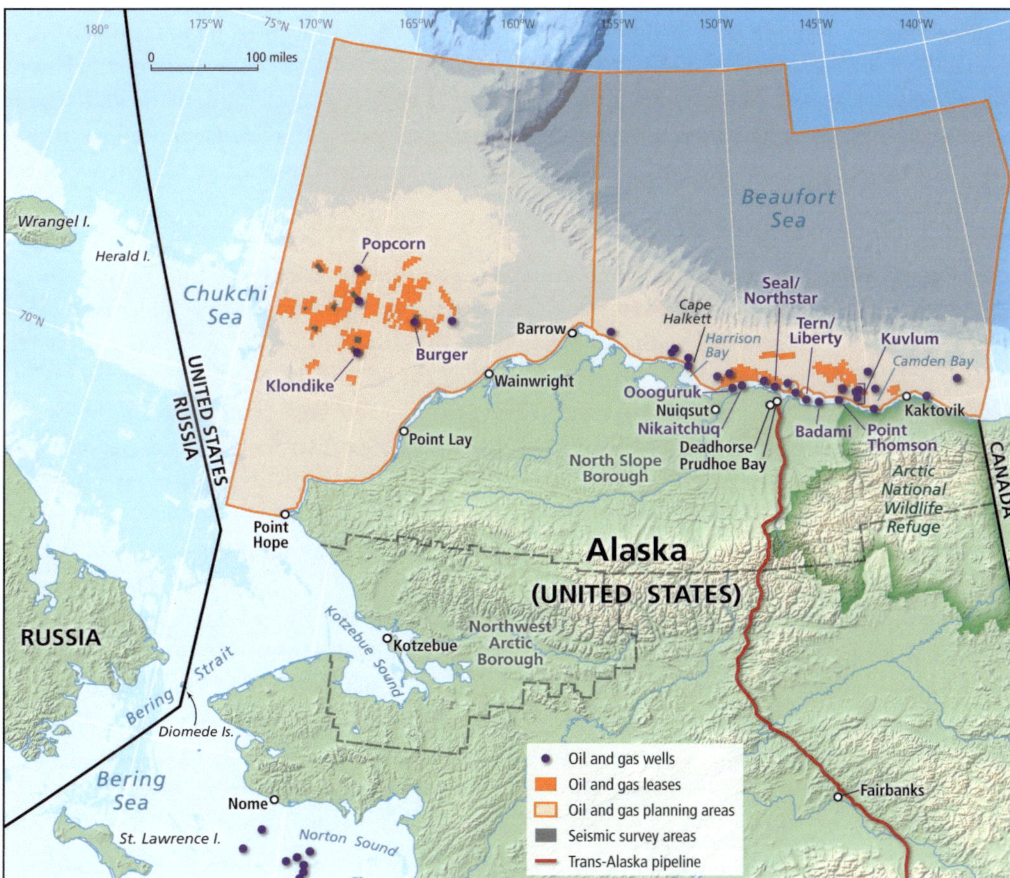

Figure 1.3 Oil and gas planning areas in the Chukchi and Beaufort Seas. Oil and gas lease areas are shown in orange, with seismic survey areas shown in gray. Selected oil and gas wells, some in Alaskan state waters and some in federal waters, are shown as purple dots. Some coastal communities and cities are also shown.

region, most of the resupply and support vessel traffic is between the offshore and points of origin of the exploration vessels (e.g., Dutch Harbor, and/or Nome). Crew change-out and some resupply also take place through Wainwright and Barrow. In the Beaufort Sea, oil exploration is primarily located between Kaktovik and Cape Halkett, especially near Camden Bay, the Colville River Delta, and Harrison Bay. Beaufort Sea exploration is closer to shore (~15-30 km), between Kaktovik and Nuiqsut. As with the Chukchi exploration effort, vessels arrive in the Beaufort Sea by traversing the Bering Strait and Chukchi Sea and rounding Point Barrow or, to a lesser degree, via Canadian waters to the east. Additional material resupply, crew change-out, and other vessel traffic are routed between the exploration areas and Prudhoe Bay.

Most of the existing Alaskan North Slope oil production infrastructure is located within a 120×40 km area between the Sagavanirktok and Colville Rivers along the central Beaufort Sea coastline. Pipelines extend approximately 30 km farther to the east, to BP's Badami and ExxonMobil's Point Thomson projects. Of the oil fields in this area, the Northstar, Oooguruk, and Nikaitchuq fields are produced from offshore facilities, with buried pipelines transmitting oil to facilities onshore. Other facilities are along the coast or in nearshore waters connected by a causeway (e.g., the Endicott Development, which was built on an artificial island). The location of these fields and pipelines influences the risk faced by communities and biological resources. Many villages are located along the coast, in large part because these areas are used by a variety of important subsistence species, especially marine mammals and birds.

Exploration drilling in the U.S. Beaufort and Chukchi Seas OCS began in the early 1980s, with 20 exploratory wells drilled between 1980 and 1989. The first discovery of oil in the Beaufort Sea OCS came in 1983 at the Tern (Liberty) field, and the largest discovery to date was made in 1993 at the Kuvlum field. Northstar was the first field to be developed in federal waters in the Beaufort Sea (although the development area spans state and federal waters), beginning production in 2001 (Holland-Bartels and Pierce, 2011). To date, 30 exploration wells[5] have been drilled in the Beaufort Sea OCS; only three of those were drilled since the mid-1990s.[6] In the Chukchi Sea OCS, five exploratory wells have been drilled.[7] All were drilled between 1989 and 1991, and several discovered hydrocarbons (USDOI, 2013).

Between 2008 and 2011, there were delays in additional exploration. In 2010, following the *Deepwater Horizon* oil spill (see Box 1.1), Secretary of the Interior Ken Salazar suspended proposed exploratory oil drilling in the Arctic.[8] In 2011, the Bureau of Ocean Energy Management (BOEM) conditionally approved Shell's 2012 Exploration Plans for both the Beaufort[9] and Chukchi[10] Seas, and the Bureau of Safety and Environmental Enforcement (BSEE) permitted Shell to begin drilling

[5] See http://www.boem.gov/uploadedFiles/BOEM/About_BOEM/BOEM_Regions/Alaska_Region/Historical_Data/OCS%20Wells%20Drilled%20by%20Planning%20Area%20-%20AK.pdf.

[6] See http://www.boem.gov/uploadedFiles/BOEM/About_BOEM/BOEM_Regions/Alaska_Region/Historical_Data/Exploration%20Wells%20Beaufort%20Sea.pdf.

[7] See http://www.boem.gov/uploadedFiles/BOEM/About_BOEM/BOEM_Regions/Alaska_Region/Historical_Data/Exploration%20Wells%20Chukchi%20Sea.pdf.

[8] See http://www.doi.gov/news/pressreleases/Salazar-Calls-for-New-Safety-Measures-for-Offshore-Oil-and-Gas-Operations-Orders-Six-Month-Moratorium-on-Deepwater-Drilling.cfm.

[9] See http://www.boem.gov/BOEM-Newsroom/Press-Releases/2011/press0804a.aspx.

[10] See http://www.boem.gov/BOEM-Newsroom/Press-Releases/2011/press12162011.aspx.

in 2012. Drilling was limited to two surface holes, due to issues with readiness of their dome containment device as well as the lack of a fully compliant spill response barge. Despite further hurdles, including the grounding of the *Kulluk* drilling unit (see Box 1.1), Shell continued to pursue drilling activity in early 2013, focusing their drilling activities at the Burger prospect in the Chukchi Sea and near the Kuvlum and Hammerhead fields in the Beaufort Sea. In February 2013, Shell announced that it was halting further exploration activities until 2014, at which point it planned to drill in the Chukchi Sea only. In its most recent announcement on January 30, 2014, Shell again postponed its drilling activities, citing as its reason a decision by the 9th U.S. Circuit Court of Appeals[11] regarding a flawed environmental impact statement for lease sales in the Chukchi Sea.

Drilling operations in Arctic waters have been and continue to be highly controversial. The National Commission on the BP Deepwater Horizon Oil Spill (2011) noted that "[t]he stakes for drilling in the U.S. Arctic are raised by the richness of its ecosystems." Many individuals and conservation organizations advocate a halt to drilling in the region. Their concerns are primarily centered around inadequate baseline and monitoring data, especially for sensitive and important ecological areas; limited infrastructure available to address oil spills; challenges presented by little daylight in winter, rough weather, sea ice, and remoteness; and a lack of effective methods for responding to oil spills (Oceana, 2008; WWF, 2010; Pew Charitable Trust, 2013). Some of these concerns are based on experiences and environmental impacts from previous oil spills in the region, notably the *Exxon Valdez* accident (discussed in Box 1.1).

Box 1.1 briefly describes four oil spills and marine transportation accidents that have relevance for the U.S. Arctic—the *Exxon Valdez*, *Deepwater Horizon*, *Kulluk*, and *Selendang Ayu*. These incidents are referenced throughout the report.

Large commercial vessel traffic through the Bering Strait to Alaska's northern regions has typically been dedicated to servicing the nearshore and onshore oil production facilities on the North Slope, transporting zinc and lead from the Red Dog mine through its port on the Chukchi Sea, and delivering fuel, equipment, and supplies to coastal communities (Figure 1.3). However, the vessel traffic situation in the region has recently changed noticeably. There has been an increase in seasonal maritime traffic from increased oil and gas exploration, ship-based oceanographic research missions from a variety of nations (including some that are newer to Arctic research, such as South Korea and China), tourism vessels, and shipping of oil and other commodities from Russia through the Northern Sea Route (Arctic Council, 2009). These trends are expected to continue, with additional traffic potential from the development of large-scale mineral deposits in the Canadian and Russian Arctic and the development of new oil fields in the Alaskan OCS. More than 300 vessels transited the Bering Strait in 2012, up from approximately 260 in 2009, according to Automatic Identification System data.[12] Of these, bulk carriers, tugs and barges, and research vessels constitute the largest categories.

---

[11] Native Village of Point Hope v. Jewell, No. 12-35287 (9th Cir. filed Jan. 22, 2014).

[12] Marine Exchange of Alaska, *Record of Recorded Transits, Bering Strait, 2009, 2010, 2011, 2012*.

BOX 1.1
Select Oil Spills and Maritime Accidents of Interest

Exxon Valdez — On March 24, 1989, *Exxon Valdez,* an oil tanker headed for Long Beach, California, struck a reef in Prince William Sound, Alaska. The collision with the reef punctured 8 of the tanker's 11 cargo tanks. The damaged tanker spilled approximately 10.8 million gallons of North Slope crude oil. It was estimated to be carrying approximately 53 million gallons when it was wrecked. At the time, it was the largest single oil spill in U.S. coastal waters. Oil from the spill reached nearly 2,100 km of coastline, approximately 200 of which were considered heavily oiled. The other 1,750 km were either lightly or very lightly oiled. The *Exxon Valdez* Oil Spill Trustee Council estimates as many as 250,000 seabirds, 2,800 sea otters, 300 harbor seals, 250 bald eagles, and 22 killer whales died as a result of the incident.

Deepwater Horizon — On April 20, 2010, there was an explosion on the *Deepwater Horizon* (*DWH*) oil platform as it drilled the Macondo Well in the Gulf of Mexico. The uncontrolled oil flow from the wellhead led to a release of about 205 million gallons at a depth of ~1,500 m. The *DWH* oil spill is, to date, the largest offshore oil spill in U.S. history. Although a final Natural Resource Damage Assessment has not yet been released, preliminary status updates provide some insight into the damage caused by the spill. Approximately 1,750 km of coastline were oiled, 220 of which were heavily oiled. As of April 2012, field teams had collected 8,567 live and dead birds, of which 1,423 were rehabilitated and released. They collected 536 live sea turtles, of which 469 were later released, and 613 dead sea turtles. An Unusual Mortality Event for cetaceans in the northern Gulf of Mexico was declared prior to the spill, in February 2010, and is still in effect.

Kulluk — On December 27, 2012, the *Kulluk,* Shell's conical drilling unit, was being towed from Dutch Harbor after drilling in the Beaufort Sea to Seattle for maintenance when its tow connection to the *Aiviq* separated. Although an emergency tow line was established between the *Kulluk* and both the *Aiviq* and the *Nanuq*, the *Kulluk*'s connections with both ships ultimately separated again on December 30. On December 31, the *Kulluk* grounded off of Sitkalidak Island, Alaska, due to strong winds and rough seas. On January 2, 2013, a salvage assessment team noted that the *Kulluk* had sustained some damage, but that it was stable and no sheen was visible. At the time that it grounded, the *Kulluk* was carrying approximately 139,000 gallons of ultra-low-sulfur diesel in addition

COMMITTEE PROCESS

The National Research Council's Committee on Responding to Oil Spills in Arctic Marine Environments was formed to address the Statement of Task in Box 1.2. The committee's work was sponsored by eight agencies and organizations—the American Petroleum Institute, the U.S. Arctic Research Commission, BSEE, BOEM, the U.S. Coast Guard, the Marine Mammal Commission, the National Oceanic and Atmospheric Administration, and the Oil Spill Recovery Institute. The committee represents a broad spectrum of knowledge and expertise related to oil spill response or the Arctic environment; committee membership is listed in Appendix A. Rosters for each of the National Research Council boards involved in the study are in Appendix B.

Committee meetings were scheduled in different locations to afford opportunities for participato the 12,000 gallons of combined lube oil and hydraulic fluid needed for onboard equipment. Approximately 316 gallons of ultra-low-sulfur diesel fuel were released from the *Kulluk's* lifeboats. On January 6, the *Kulluk* was refloated and moved to nearby Kiliuda Bay. Multiple salmon-bearing streams are located near the grounding site and Kiliuda Bay. Sitkalidak Island and Kiliuda Bay are within the area designated as critical habitat for Steller sea lion and southwest sea otter populations. These areas also provide habitat for waterfowl and shorebirds (including the Endangered Species Act-listed Steller's eider), as well as harbor seals. Fin and humpback whales may also have been in Kiliuda Bay. However, no specific impacts to wildlife have been reported.

Selendang Ayu — On November 28, 2004, the Malaysian-registered bulk freighter *Selendang Ayu* departed Seattle, Washington, and began its trip to Xiamen, China. The vessel was loaded with soybeans and 1,000 metric tons of fuel. On December 6, in the Bering Sea, the vessel experienced engine failure. Despite attempts to fix the engine, it would not restart. Efforts over the next 2 days to tow the vessel were compromised and ultimately failed, due largely to weather. After drifting significantly, the *Seledang Ayu* ran aground near Unalaska Island, spilling 336,000 gallons of fuel and diesel fuel. Resources that were reportedly damaged from the spill include birds, fish, and vegetation. Following the incident, shoreline cleanup assessment teams identified nearly 115 km worth of shoreline segments that would require additional treatment. Some of the most heavily oiled areas were beaches located at the mouths of streams, which serve as habitat for anadromous fish. The carcasses of approximately 1,700 birds were either recovered or documented.

SOURCES: *Exxon Valdez* Oil Spill Trustee Council (http://www.evostc.state.ak.us/); NOAA Office of Response and Restoration (http://response.restoration.noaa.gov/exxonvaldez); NOAA Gulf Spill Restoration (2012); Alaska Department of Environmental Conservation (http://dec.alaska.gov/spar/perp/response/sum_fy13/121227201/jic/TimelineAndMap.pdf; http://dec.alaska.gov/spar/perp/response/sum_fy13/121227201/121227201_sr_09.pdf; http://dec.alaska.gov/spar/perp/response/sum_fy13/121227201/121227201_sr_13.pdf); *Kulluk* Tow Incident website (https://www.piersystem.com/go/doc/5507/1674235/#.UoOk8-LO2M0); National Transportation Safety Board marine accident brief (http://www.ntsb.gov/doclib/reports/2006/MAB0601.pdf); U.S. Fish and Wildlife Service (http://www.fws.gov/alaska/fisheries/contaminants/spill/sa_injury.htm).

tion by committee members, stakeholders, interested parties, and the public. Prior to formal committee meetings, committee and staff members attended workshops in Kotzebue and Barrow in order to interact with local stakeholders and decision makers. Committee activities included information gathering through presentations and discussion, small group meetings, and writing sessions. The committee drew on the large volume of recent reports that deal with many aspects of the Arctic, especially those related to ecosystem change, developments in shipping and energy exploration, and policy. They also examined experiences and lessons learned from several oil spills and marine transportation accidents. The committee notes that the report sponsors asked for a specific focus on U.S. Arctic waters of the Bering Strait, Chukchi Sea, and Beaufort Sea. This region is the main focus of the following chapters, with only limited emphasis on U.S. Arctic waters in the Bering Sea and points farther south for context. There is also limited focus on international Arctic waters outside U.S. jurisdiction, except as it pertains to oil that could potentially spread into U.S. waters.

Box 1.2
Statement of Task

The National Research Council (NRC) will assess the current state of science and engineering regarding oil spill response and environmental assessment in the Arctic region (with a specific focus on the Bering Strait and north), with emphasis on potential impacts in U.S. waters. As part of its report, the NRC-appointed committee will assess existing decision tools and approaches that utilize a variety of spill response technologies under the types of conditions and spill scenarios encountered at high latitudes. The report will also review new and ongoing research activities (in both the public and private sectors), identify opportunities and constraints for advancing oil spill research, describe promising new concepts and technologies for improving the response, including containment (surface and subsurface) approaches to reduce spill volume and/or spatial extent, and recommend strategies to advance research and address information gaps. The committee will also assess the types of baselines needed in the near term for monitoring the impacts of an oil spill and for developing plans for recovery and restoration following an oil spill in U.S or international waters where a spill could potentially impact U.S. natural resources. For assessing the state of the science, the committee will address the following topics:

(1) Scenarios. Identify areas in U.S. or adjacent waters where current or potential activities could lead to an oil spill in the marine environment (marine transportation routes, cruise ships, fishing, pipeline locations, fuel storage facilities, oil and gas exploration and production). The scenarios would include descriptions of oil type (including biofuels and diesel fuel) and possible volume and trajectories of spills, season, and geographic location, including proximity to local communities and highly valued fish, bird, and marine mammal habitats.

(2) Preparedness.

- Describe the anticipated operating conditions, such as ice conditions, currents, prevailing winds, weather, amount of daylight, sea state, and distance/accessibility from responders and resources. This will include an evaluation of the state of hydrographic and charting data for higher risk areas.
- Assess infrastructure (including communication networks), manpower, and training necessary to operate in these conditions.
- Identify avenues for participation of and communication with indigenous communities and regional governmental (e.g., Alaska State) entities during planning and response.

REPORT ORGANIZATION

While the following chapters are roughly divided along the charges within the Statement of Task, there are many areas where the charges are woven together or split apart. Chapter 2 discusses the environmental conditions, baseline needs, and marine activities in the U.S. Arctic. This chapter mainly addresses task 4 on strategies for establishing the types of baseline information needed before a spill, but also addresses part of task 2 regarding anticipated operating conditions. Chapter 3 examines current and ongoing oil spill response research, as required in task 3, to evaluate existing

- Build on existing agreements and identify gaps for international cooperation in establishing locations for incident command management, staffing, and supplying oil spill response infrastructure, recognizing the international interests in navigation and resource exploitation in Arctic environments.

(3) Response and Cleanup. Evaluate the effectiveness and limitations of current methodologies used in response to a spill in Arctic conditions.

- Assess utility of existing and promising new technologies to detect, map, track, and project trajectories of spills under the anticipated operating conditions (e.g., ice conditions, visibility).
- Evaluate the effectiveness of oil spill response technologies under the following criteria:
    - Operation under various conditions and time frames (volatile fractions, wind, sea state, temperature, degree of emulsion, oil type and viscosity);
    - Spatial and temporal dimensions of the spill and the response;
    - Transportation of equipment to remote areas;
    - Natural oil degradation rates; and
    - Ancillary effects of response operations on the indigenous communities, environment, and marine species.
- Assess the potential response strategies for the separation and recovery of oil from marine waters, on or associated with ice, sediments, and the shore zone, including an assessment of their contributions toward habitat recovery. This assessment will include discussion of constraints in the handling, storing, and disposing of recovered oil in situ or in remote locations, the volume of material to be treated, selection of methodologies for incineration or recycling onboard ship or in a remote location, and the further disposal or transport of the recovered product. The assessment will also include discussion of fate and effects of unrecovered oil left to biodegrade and weather in Arctic environments.
- Assess the capabilities and constraints for minimizing impacts and enhancing recovery of wildlife through deterrence and rehabilitation.

(4) Strategies for Establishing Environmental Baselines for Spill Response Decisions. Characterize the types of baseline information needed in the event of an oil spill. Evaluate existing pre-spill strategies for resource protection and identify additional protection options for resources at risk. Identify sampling and monitoring priorities for establishing baseline conditions and evaluating impacts of a potential spill.

and new technologies available to respond to an oil spill and their effectiveness. Chapter 4 addresses operations, logistics, and coordination in response to an Arctic oil spill, which relates to task 2 on operating conditions and infrastructure, as well as the opportunities to build on existing agreements and partnerships. Chapter 5 deals with evaluation of response technologies and strategies, an element of task 3. To address task 1, the committee introduces some representative scenarios as part of the risk framework discussion in Chapter 3, with further scenario development and discussion in Chapters 4 and 5.

2 Environmental Conditions and Natural Resources in the U.S. Arctic

The components of the Arctic system interact with each other in a complex, evolving pattern. This chapter provides an overview of the physical and biological ocean processes and environments of the Bering Strait and the Chukchi and Beaufort Seas. This is important for understanding current conditions, but also for understanding trends, changes, and future data needs. This knowledge is essential to support safe operations in the Arctic marine environment; to guide oil spill prevention, response, and restoration; and to prioritize sampling and monitoring needs.

The chapter begins with a discussion of the physical environment: ocean, marine weather, sea ice, and coastal characteristics and processes—conditions that would be encountered in the event of an oil spill in Arctic waters. A discussion of Arctic ecosystems follows, with an emphasis on biological information that would be important for oil spill response, including monitoring approaches and data needs for incorporation into spill trajectory models or ecosystem models. A final section discusses current research and monitoring as well as additional needs to advance understanding of the Arctic system.

OCEAN PROCESSES AND CHARACTERISTICS

Three principal branches of Pacific origin water (Alaska Coastal Water, Bering Shelf Water, and Anadyr Water; Figure 2.1) enter the Bering Strait and circulate through the Chukchi and Beaufort Seas. The northward transport of water through the Bering Strait is principally driven by large scale sea level differences between the Pacific and Arctic Oceans, and opposes the prevailing winds; variability in the currents is predominantly wind driven (Weingartner et al., 2005). The summer retreat of ice in this sector of the Arctic has been linked in part to warm Bering Strait inflows and flow pathways (Woodgate et al., 2010).

Water properties in the Chukchi Sea are set by processes of sea ice melt and growth, winds, and inflows of river water and Pacific Ocean via the Bering Strait, as well as from the East Siberian Sea. Transport of water from the Chukchi and shelf regions to the deep Canada Basin takes place mainly through the Barrow and Herald canyons (Weingartner et al., 2005). Currents that flow eastward along the Beaufort Sea continental slope generate eddies that can propagate into the basin interior,

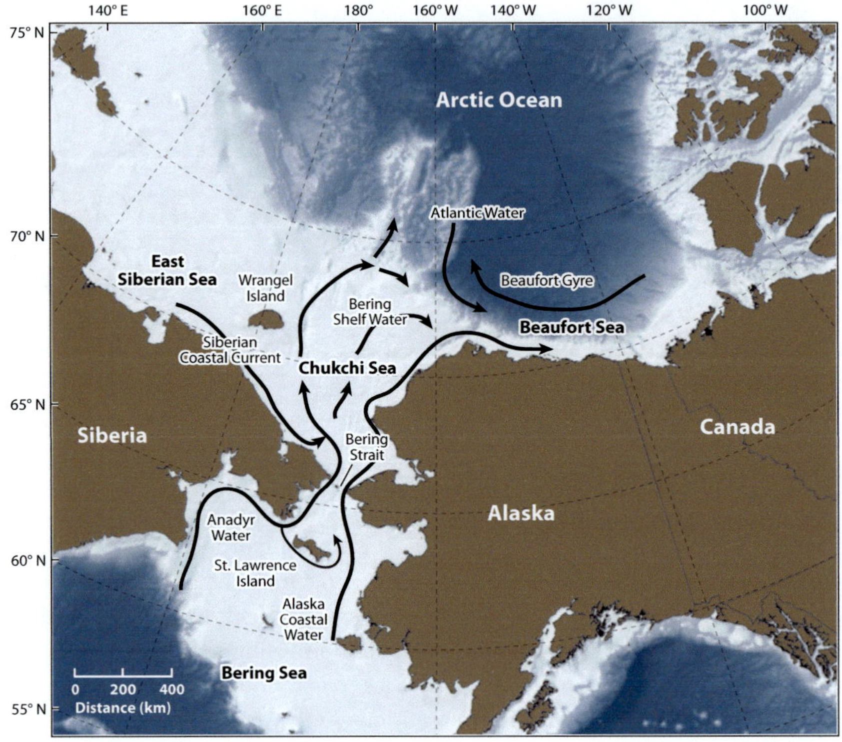

Figure 2.1 Water masses and sea ice extent in Bering Strait and Chukchi and Beaufort Seas. SOURCE: Grebmeier (2012).

and surface wind forcing can also drive water offshore (e.g., Pickart et al., 2005). Transport pathways, seasonal evolution, and mixing of the water masses are outlined in Weingartner et al. (2013b) and references therein.

Weingartner et al. (2005) found that mean flows over the Chukchi shelf are generally less than 10 km/day except in Barrow Canyon, where maximum current speeds can briefly reach 85 km/day. Elsewhere in the Chukchi Sea, maximum currents are between approximately 25 and 40 km/day. The mean flow is north-northeastward over the central Chukchi shelf, opposite to the mean winds. Weingartner et al. (2013a) examined in detail winds and water properties over the central portion of the northeastern Chukchi Sea shelf (40-45 m depth) in summer and fall 2008-2010. They showed that temperature and salinity properties can vary significantly over only a few tens of to a hundred kilometers. Along the shoreline of the Beaufort Sea in open water or loose ice conditions (July to mid-October), surface currents are predominantly wind driven (speeds typically exceed 8 km/day and maximum currents can reach almost 80 km/day, while in winter under fast ice, currents are weak—less than ~2.5 km/day—and forced predominantly by tides) (Okkonen and Weingartner, 2003). Of utmost importance to oil spill response is the rapid variability of the wind-forced surface ocean circulation. High-frequency radar systems in the Chukchi Sea, which map surface ocean currents, indicate complex flow patterns that can reverse direction in a matter of hours and can vary significantly in both magnitude (0-85 km/day) and direction over spatial scales of less than 10 km.

The Arctic Ocean is strongly stratified, with a fresh, low-density mixed layer up to 40 m deep in the Beaufort Sea. In the summer, very fresh, warm mixed layers only a few meters deep are observed in parts of the Chukchi and Beaufort Seas (e.g., Toole et al., 2010; Weingartner et al., 2013b). This stratification has important implications for the pathways and fate of spilled oil. In the *Deepwater Horizon* spill, oil plumes rising through the stratified ocean water column spread out at some level of neutral buoyancy and became trapped at depth (Socolofsky et al., 2011). The impact of the strong Arctic Ocean stratification on the subsurface evolution of spilled oil, particularly when surfactants are used (e.g., Adalsteinsson et al., 2013), is an important response planning consideration, as oil that is trapped at depth will not be transported by surface circulation. For this reason, there is a need for characterization and monitoring of Beaufort and Chukchi subsurface circulation, which can be as complex as the surface flow and can be in opposing directions (e.g., Weingartner et al., 1998; Proshutinsky et al., 2009; Morison et al., 2012).

Contaminants residing either in the surface or in the subsurface ocean are subject to redistribution by coastal ocean upwelling and downwelling events; such events have been well studied in the Beaufort Sea (see Williams et al., 2006; Schulze and Pickart, 2012; Pickart et al., 2013). Strong easterly winds have been observed to bring warm, salty deep water to shallow depths along the Beaufort Sea continental slope, with increased frequency of upwelling events in the absence of concentrated pack ice (Pickart et al., 2009). Similarly, downwelling events forced by westerly winds cause the descent of near-surface waters along the coast (Dunton et al., 2006).

Ocean eddies are common in both the Chukchi Sea and the Beaufort Sea (e.g., Manley and Hunkins, 1985; Pickart et al., 2005; Timmermans et al., 2008). Eddies centered at depths ranging from a few tens to hundreds of meters (with horizontal scales from a few kilometers to tens of kilometers) can trap and transport packets of water, or (in the case of a spill) entrained oil, over hundreds of kilometers (Provenzale, 1999; Haller and Beron-Vera, 2013). Satellite measurements reveal that the surface distribution of oil in the *Deepwater Horizon* spill was influenced by eddies in the Gulf of Mexico, which can extend to 800 m depth (Walker et al., 2011). In addition to larger-scale eddies, there is complicated smaller-scale flow structure (characterized by horizontal scales around 1 km or less) in the ocean mixed layer beneath sea ice in the Beaufort Sea (Timmermans et al., 2012) and in the mixed layer in ice-free conditions in the Chukchi Sea (Timmermans and Winsor, 2013). This small-scale flow field, which is characterized by strong convergence and divergence zones, has been shown to have an important influence on tracer distribution patterns in midlatitude, ice-free regions (Zhong and Bracco, 2013).

Ocean storm surges related to persistent high winds are an important factor for consideration in coastal spill response. Loss of Arctic sea ice has been shown to increase storm surge frequency (Lesack and Marsh, 2007; Pisaric et al., 2011). Extreme coastal flooding from water forced onshore by winds has been documented along the Canadian Beaufort Sea coast (see, e.g., Harper et al., 1988, who show maximum storm surge elevations of 2.5 m above mean sea level recorded at Tuktoyuktuk, Northwest Territories, Canada). These storm surges move ocean water into low-lying coastal environments, bringing salt and contaminants (in the event of a spill) that can have negative impacts on nearshore and terrestrial ecosystems (e.g., Thienpont et al., 2012).

MARINE WEATHER AND SEA ICE PROCESSES

Marine Weather

There are a number of key weather parameters in the Beaufort and Chukchi region that can affect oil spill response, including air and water temperature, winds, low visibility, and hours of daylight. These conditions were highlighted as challenges to oil and gas operations and scientific research in the Arctic by the National Commission on the BP Deepwater Horizon Oil Spill (2011), among others. Figure 2.2 shows a temperature record collected at Barrow over a 5-year period from 1999 to 2005 (Szymoniak and Devine, 2006). Air temperatures are low through most of the year and exhibit little variability from year to year (Figure 2.2).

Stegall and Zhang (2012) analyzed three-hourly North American Regional Reanalysis winds in an in-depth review of wind statistics in the Chukchi–Beaufort Seas and Alaska North Slope region for the period 1979-2009. They found a distinct seasonal cycle, with lowest wind speeds (~2-4 m/s) in May and largest (~9 m/s) in October, with extreme winds (up to 15 m/s) that are most often found in October. An increasing trend in the frequency and intensity of extreme wind events was identified over their study period; 95th percentile winds in October increased from 7 m/s in 1979 to 10.5 m/s in 2009. Wind fields over offshore areas are not always well-captured by coastal station data, which comprise the majority of source data for reanalysis winds. For example, along the North Slope, the significant influence of the Brooks Range in winter and the sea breeze effect from thermal gradients between land, ocean, and ice in summer can lead to stark differences between the coastal and offshore wind regimes. Wind measurements from Pelly Island in the Canadian Beaufort Sea, which may better represent the stronger and more variable Beaufort Sea marine winds than coastal stations to the west (see Manson and Solomon, 2007), recorded peak wind speeds of more than 20 m/s in most months in the period 1994-2008 (Fissel et al., 2009). Wind speed distribution can be used to assess how often a spill response technique such as in situ burning could be used. For example, a general upper wind limit for successful ignition and effective burning in booms or in situ burning is on the order of 10 m/s (Buist et al., 1994).

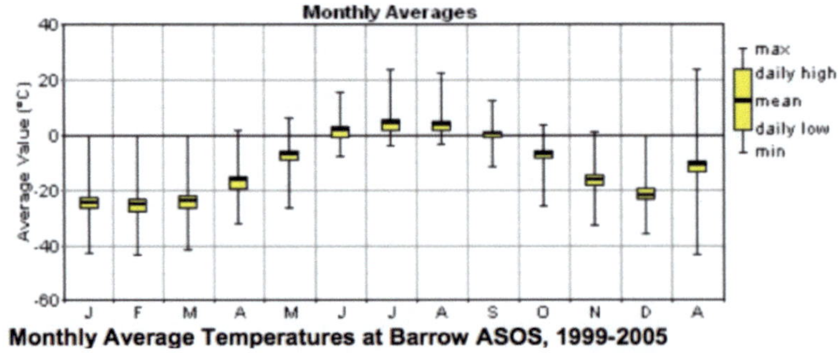

Figure 2.2 Monthly average temperatures at the Barrow Automated Surface Observing System (ASOS) from 1999-2005. SOURCE: Szymoniak and Devine (2006).

Limited daylight can be a major issue for oil spill response during freeze-up and over the winter. Off the Beaufort coast, the maximum of 21 hours of daylight during the breakup season in August reduces to an average of 11 hours in October. However, in practice, twilight can increase the available operational time beyond the hours of sunrise to sunset. From late November to January there is no daylight. Low-visibility conditions in the Beaufort Sea offshore occur most often during the breakup period in July and August. In contrast, the freeze-up period in October has less likelihood of low visibility (Dickins et al., 2000).

The Beaufort Sea wave environment can present a significant challenge to oil spill response. Waves are predominantly generated during the open water season and generally propagate from the east and northeast, although recent analyses suggest sizeable waves now also come from the west (Fissel et al., 2012). For much of the summer (July to August), the close proximity of sea ice is thought to prevent high sea states from forming. However, since 2001, upward-looking sonar measurements in the Beaufort Sea have shown a trend of large waves being present in summer and fall for longer durations, with significant interannual variability in wave heights (Fissel et al., 2012). It has been hypothesized that because of larger fetches in summer, the summer wave field now contributes significantly to a marginal ice zone of broken-up floes along the Beaufort Sea ice edge. After the initial freeze-up in October, wave heights become limited. During open water periods, maximum sea states can be estimated using the Beaufort scale relationship.

The Meteorological Service of Canada has provided a wind and wave hindcast of the Beaufort Sea, covering a 41-year period between January 1970 and December 2010[1] (see also an earlier study by Eid and Cardone, 1992). The maximum significant wave height over this period in the southern Beaufort Sea was ~9 m, with mean significant wave heights ~1-2 m. Francis et al. (2011) examined significant wave heights in the Chukchi Sea during 1993-2011 and found a 2-2.5 cm/yr increase in significant wave height, which is consistent with increased fetch accompanying sea ice retreat in this region over the same period. It is of note that about 80 km offshore there is good correlation between wave heights determined from moored measurements and satellite observations in the southeast Chukchi Sea, with worse agreement closer than ~10 km to the coastline.

Depending on the time of year, different sets of operating limits can cause interruptions to marine and air activities. From December to March, sea state is not an important factor. Operational downtime is dominated by darkness, snow, and low temperatures. Sea state and temperature are not critical factors from March to June or July; instead, downtime is related to wind and visibility such as fog and low clouds. From August to October, sea state is an important factor, while low air temperature and increasing darkness become critical from late October onward. Box 2.1 illustrates some examples of risk factors in the marine operating environment and demonstrates ways weather parameters can have an impact on marine operations and response.

Sea Ice

Sea ice is a critically important component of the Arctic marine environment, and understanding the ice environment is essential to anticipating the likely behavior of oil in, under, and on ice and

[1] See http://www.oceanweather.net/msc50waveatlas.

BOX 2.1
Examples of Risks Associated with Oil Spill Response Due to Weather Conditions

Adverse weather conditions can have impacts on the feasibility of oil spill response, especially in relation to marine and airborne operations in support of the response. This box provides examples of risks associated with particular weather conditions. Actual operating limits are determined by an operator for each installation and piece of equipment.

- Sustained wind speeds greater than 25 knots (~13 m/s) could
    - Hinder crane operations and equipment use on response vessels, with a possibility of swinging or uncontrollable loads;
    - Limit in situ burning, as a typical wind threshold for successful burn operations is 20 kt (~10 m/s) or less;
    - Limit surface dispersant application from vessels and aircraft;
    - Limit mechanical recovery operations, such as skimmer deployment and boom containment;
    - Hamper small boat operations due to the potential for severe sea states, breaking waves, and superstructure icing; and
    - Hinder helicopter approach and landing on offshore helidecks.

- Sea states greater than 1-1.5 m could
    - Limit boom effectiveness, as wave overtopping leads to loss of contained oil;
    - Impede small boat operation, due to waves, wind, and icing potential;
    - Contribute to seasickness and/or fatigue, impacting personal safety and effectiveness; and
    - Jeopardize safety on deck from slippery and icy surfaces.

- Visibility that is less than visual flight rules or instrument flight rule minimums (due to weather or season) could
    - Limit helicopter landings when cloud ceilings or visibility are below minimum standards set by the Federal Aviation Administration or company policy; and
    - Curtail aerial dispersant spraying;
    - Limit oil spill monitoring by preventing direct visual observations.

- Extreme cold temperatures (less than −35°C) could
    - Impact safety on deck, due to effects from wind chill;
    - Impact responder safety because of potential for frostbite;
    - Decrease worker efficiency from fatigue, leading to a need for frequent rest breaks;
    - Contribute to equipment breakdown due to changes in oil viscosity, hydraulic leaks, or mechanical failure; and
    - Limit helicopter operations, which have a lowest acceptable operating temperature set by operators and manufacturers.

determining applicable response strategies. At present, marine operations in the northern Chukchi and Beaufort Seas generally take place from late July to September, but future developments could lead to extended operating seasons or even year-round offshore oil production. Even in the summer, ice can intrude on drilling locations and shipping routes. Furthermore, ice-free regions can transition to ice-covered conditions in a matter of days at the start of fall freeze-up. Sea ice therefore needs to be considered as a possible operating condition for spill response planning. In areas that are of interest to industry, sea ice monitoring is generally performed more consistently and with higher resolution than elsewhere in the Arctic.

This section describes seasonal and spatial characteristics of the Beaufort and Chukchi Sea ice environments in order to set the context for spill response procedures and obstacles in the next chapter. For additional background, the National Snow and Ice Data Center (NSIDC) provides concise descriptions of many key processes in ice formation and decay, as well as the different forms of sea ice[2] (recent sea ice conditions are also summarized in Perovich et al., 2013). Worldwide standards for sea ice charting and reporting are maintained by the Joint World Meteorological Organization-Intergovernmental Oceanographic Commission Technical Commission for Oceanography and Marine Meteorology, while the Arctic and Antarctic Research Institute in Russia maintains the current web-based edition of "World Meteorological Organization WMO-IOC Sea Ice Nomenclature No. 259." *Ice concentration*, the areal extent of ice relative to open water, is expressed as tenths of ice coverage (e.g., 1/10 = 10% coverage of ice by area).

Sea ice has a complicated seasonal evolution that is a function of seasonal temperature variations and mechanical forcing; its structure and evolution differ significantly from the coastal zone to offshore. *Land-fast* ice refers to sea ice that is frozen along the shore, partially frozen to the seabed in shallow water (less than 2 m), and largely free-floating in deeper water (typically 15-30 m), where grounded ridges can act to anchor the sheet against drifting pack ice forces. Land-fast ice is most extensive along broad, shallow shelves. Although fast ice along the Beaufort Coast is generally stable near shore after December, severe storm events can cause winter shearing and movement and breakaway events, where large sections drift away from the fast ice edge in deeper water. These events are more common off parts of the Chukchi coast. Beyond the bottom-grounded land-fast ice zone, floating fast ice extends seaward as the season progresses, until it reaches an outer limit within a shear zone (Eicken et al., 2006; Figure 2.3); this zone of often significantly deformed ice can be highly variable in extent but typically occurs between the 15- and 25-m isobaths (also referred to as the *Stamukhi* zone in traditional Alaskan references such as Kovacs [1976] and Reimnitz and Kempema [1984]). The stable and relatively smooth nearshore areas of land-fast ice (out to approximately 12 m water depth) are used in the Beaufort Sea along the North Slope to construct winter ice roads that routinely carry heavy equipment in midwinter (Potter et al., 1981; Masterson, 2009). Land-fast ice also serves as an important hunting and traveling platform for Arctic coastal communities, as noted by Mahoney (2012).

Drift ice floats freely on the ocean surface without any stable connection to land. *Pack ice* is drift ice whose concentrations exceed 6/10. The pack can open or close on the order of hours in response to winds and/or ocean currents (Potter et al., 2012). Typical midwinter pack ice drift rates in the

[2] See http://nsidc.org/cryosphere/seaice/index.html.

Beaufort Sea are on the order of 5 km/day (Melling and Riedel, 2004). Ice drift rates can exceed 50 km/day, based on 80th percentile exceedance values published by Melling et al. (2012) from moorings in the Canadian Beaufort Sea. Peak values measured in the same dataset over a 30-minute period reached 1.2 m/s. Even higher short-duration speeds (under 12 hours) are possible along the U.S. Beaufort Sea coast, where the mountain barrier effect of the Brooks Range amplifies offshore east-west winds. The net displacement of ice past a mid-shelf site north of Tuktoyaktuk between mid-October and mid-May was almost 2,000 km in 2007-2008. The actual distance along a drift path, including loops and backtracking, could be larger.

The distinction between fast ice and pack ice and the location of the ice edge at different times in the winter has important implications for oil fate and behavior. Ice features embedded in fast ice are generally static, so oil spilled into this stable ice regime is likely to remain very close to the discharge point (within hundreds of meters) for much of the year. In contrast, oil spilled into a pack ice environment north of the fast ice edge will drift with the ice over time (Wadhams, 1976, 1981; Wilkinson et al., 2007; Dickins, 2011).

In addition to location, the timing of a spill in relation to the sea ice seasonal cycle can control oil behavior and related response options. When sea ice first forms on the ocean surface in the fall, it transitions through a range of stages (depending on atmospheric and ocean conditions)—a growth process that eventually leads to first-year ice, which in a single season may become 1.5 to 2 m thick by late winter (April/May) in the Chukchi and Beaufort Seas (Dickins et al., 2000; Wadhams et al., 2011; Wadhams, 2012a). Oil spilled under growing first-year ice will become encapsulated in a layer of new ice within 12-48 hours, depending on the thickness of the ice, snow cover, and ambient

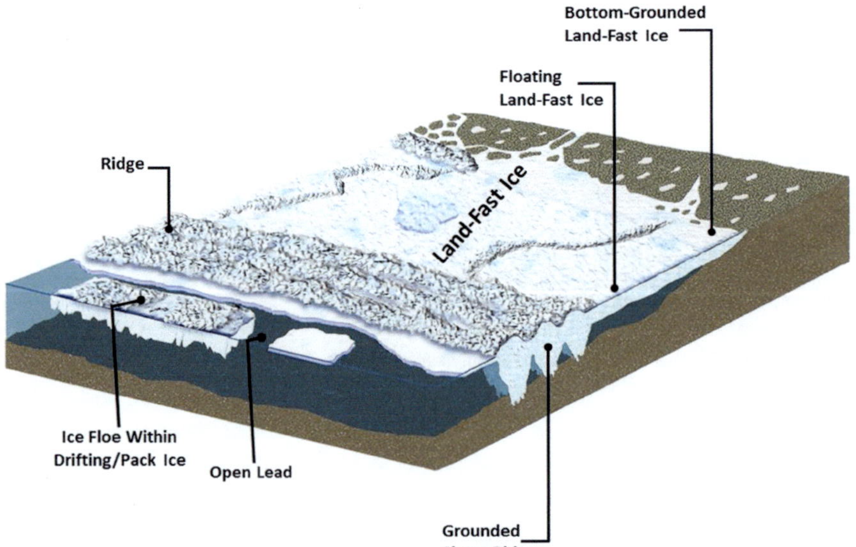

Figure 2.3 Cross-section of typical Beaufort Sea ice zones. SOURCE: Used with permission from University of Alaska Fairbanks, Geophysical Institute.

air temperature. This process has been documented in a series of large-scale field studies (Dickins and Buist, 1999; Dickins, 2011).

In most Arctic areas, the overlying snow layer usually starts to melt in late May or early June and is gone by early July. Meltwater from the snow creates a network of meltwater pools over the ice surface. In first-year ice, oil trapped under or encapsulated within the ice will migrate to the surface through channels left in the ice as the heavier brine drains out. Once on the ice surface, winds will push the oil into thicker patches on the lee side of melt pools that can be ignited and burned (NORCOR Engineering & Research Ltd., 1975; Dickins and Buist, 1981; Brandvik et al., 2006).

Much of the ice cover encountered beyond the nearshore land-fast ice zone is deformed from crushing and shearing or from young ice rafting over itself in the first few months following freeze-up, forming ridges and ice rubble. These processes can create several-meters-thick patches of ice made up of multiple thin sheets. Ridges can extend well over 30 m below the surface and 5 m or more above the surrounding ice field (Tucker et al., 1979; Wadhams and Horne, 1980; Weeks and Hibler, 2010). Monitoring ice thickness is a particularly serious technical question, as current satellite methods are deficient in this area; both CryoSat and ICESat satellite sea ice thickness data are subject to issues regarding validation. Recent progress has been made through the analysis of distributed moored sonar measurements in the Beaufort Gyre region from 2003 to 2012 (Krishfield et al., 2014). Over this period, Krishfield et al. (2014) found an increase in the fraction of ice floes less than 0.3 m thick and a reduction in older, thicker ice floes. A 0.5-m decline in mean ice draft over the 9-year period was observed; most of this decline in sea ice thickness occurred in 2007 and 2008. Even areas of smooth, level ice that have not been subject to dynamic deformations show distinct variations in thickness, with significant undulations or troughs in the underside of the sheet. This spatial variability in ice sheet thickness can increase with time as the ice grows, creating localized areas where large volumes of oil trapped under ice could be effectively contained in relatively small areas (Wilkinson et al., 2007).

Multiyear ice, which has survived one or more melt seasons, can be highly variable in thickness, with a typical maximum of 3-4 m thickness (Wadhams et al., 2011). Multiyear ice is significantly fresher than first-year ice, without a well-defined network of brine channels (e.g., Johnston, 2004). This has implications for oil migration that are not well understood, although it is generally believed that it may take several seasons for oil trapped under multiyear ice to appear on the surface. Limited field tests with actual oil spilled under multiyear floes provide inconclusive results (Comfort and Purves, 1982).

There is a vast historical body of knowledge on the ice environment in the Beaufort Sea, and to a lesser extent in the Chukchi Sea (e.g., Barry et al., 1976; 1979; Kovacs, 1976; Stringer et al., 1980; Dickins, 1984; Reimnitz and Kempema, 1984; St. Martin, 1987; Voelker and Seibold, 1990). Much of this work was completed in the late 1970s and early 1980s, at the peak of the offshore exploration boom in the region. Most reports from that time focus on the morphology and dynamics of fast ice, the shear zone, and seasonal pack ice. Wadhams (2000) and Weeks and Hibler (2010) provide overviews of sea ice knowledge.

The Chukchi Sea was the subject of intense study from 1985 to 1990, related to exploration activity and development proposals from a number of companies. In the early 1990s, interest in transporting gas from the North Slope led to a series of comprehensive studies of ice conditions in several locations, including Wainwright. Publicly available sources of modern ice information for the Chukchi Sea are limited in comparison with the Beaufort Sea, partly because of the shorter history of commercial interest. Consequently, fewer historical ice datasets are available; older references include the original Alaska Marine Ice Atlas (LaBelle et al., 1983) and studies of sea ice motion and ice ridging (Weeks et al., 1980; Pritchard and Hanzlick, 1987). Recent industry-sponsored reports are generally proprietary to the operators, although modern ice information is available from researchers at the University of Alaska Fairbanks (e.g., Eicken et al., 2006; Mahoney et al., 2007, 2012). Mahoney (2012) notes that the Chukchi and Beaufort Seas have experienced some of the most significant changes in terms of sea ice extent, thickness, and age in recent years (Maslanik et al., 2007; Kwok and Cunningham, 2010). Further discussion of projected changes in timing and extent of sea ice in the Bering and Chukchi Seas is found in Douglas (2010).

Summer ice conditions are highly variable and largely dictated by wind patterns. In the Beaufort Sea, persistent easterly winds tend to move the pack away from shore, promoting extensive clearing along the coast, while westerly winds tend to keep the pack ice close to shore and limit the extent of summer clearing (DF Dickins Associates Ltd. and OASIS Environmental, 2006). In recent summers, ice drift in the Beaufort Sea has exhibited a stronger drift component toward the North Pole, moving ice away from the coast (Hutchings and Rigor, 2012; Mahoney, 2012). Hutchings and Rigor (2012) found this to be an important factor leading to the low sea ice extent in summer 2007. The length of the melt season has increased by over 10 days per decade in the Chukchi and Beaufort Seas over the past 30 years (Markus et al., 2009). Over a 12-year period, the duration of summer open water in the central Chukchi Sea ranged from 8 to 24 weeks (Wang et al., 2012). The average duration of open water in the Chukchi has lengthened significantly on average over the past 30 years. There is also a clear gradation in open water duration with latitude following the retreat of the pack ice edge, from a historical average of 20 weeks or more off Cape Lisburne, to less than 4 weeks north of 72°N. While sea ice extent is at a minimum in the Chukchi-Beaufort region in the latter half of September, ice incursions lasting up to several weeks can occur when the remaining offshore pack ice is driven into shore by sustained westerlies or when remaining thick grounded remnants of the shear (*Stamukhi*) zone can float free in summer and drift through the region (DF Dickins Associates Ltd. and OASIS Environmental, 2006).

The fall transition from the first appearance of new ice to almost complete ice cover (8/10 or more) nearshore occurs rapidly in the Beaufort Sea, often within a week or less. Initial ice growth along the coast can reach 30 cm within two weeks after the first occurrence of new ice (Dickins et al., 2000). Farther offshore, freeze-up is characterized by the presence of substantial amounts of grease ice (thin layers of clumped crystals on the ocean surface that can resemble an oil slick) or slush before the first consolidated new ice sheet appears. The slush between thicker floes has been observed to significantly restrict oil spreading in leads, maintaining oil in patches thick enough for effective ignition and burning (Buist and Dickins, 1987).

It is important to understand how the different ice regimes develop through the winter in the event that oil remaining from an accidental release remains trapped in the ice after freeze-up. In the winter, the Beaufort Sea pack ice moves in an episodic, meandering fashion with a typical net westerly drift in response to wind and currents. Mean monthly ice speeds reach a maximum in November and December (typically 9-13 km/day) and gradually decrease as the ice pack thickens and becomes more consolidated through January and February. Mean monthly speeds reach a minimum in March and April, with typical values of 3-5 km/day, although there are long periods (weeks or more) when the offshore ice moves very little or meanders locally at low speeds with no persistent sense of direction (Melling and Riedel, 2004). Average winter ice drift speeds in the Chukchi tend to be significantly greater than in the Beaufort and can exceed 40 km/day for 24 hours or more.

Freeze-up along the Chukchi coast begins in early October off Barrow and progresses south to Cape Lisburne by late October. The offshore ice cover in the Chukchi Sea often takes much longer to consolidate, with open water stretching well into November in many years. Prevailing easterly to northeasterly winds across the northern Chukchi Sea often create an area of open water between the pack ice and the landfast ice, with young ice offshore of the grounded perimeter of the landfast ice zone along the northern Chukchi coast (Figure 2.4).

Frequent winter breakaway events can substantially alter the extent of fast ice along the Chukchi Sea coast in a matter of hours, as floes that can be several kilometers across fracture and drift out into open water stretches. In early winter, the fast ice remains unstable right into the coast until December and occupies a limited extent compared to the Beaufort Sea. In some areas—for example, north of Wainwright—the fast ice is less than 3 km in width, a function of the much steeper bottom slope and lack of broad shallow shelf area due to the presence of the upper Barrow Canyon intersecting the shelf in this region when compared to the shallower, but narrower, Beaufort Sea shelf.

Sea ice dominates the Chukchi Sea from November to early July on average, 4 to 6 weeks shorter than in the Beaufort Sea. Fast ice begins to break up in early June, a month ahead of the Beaufort Sea.

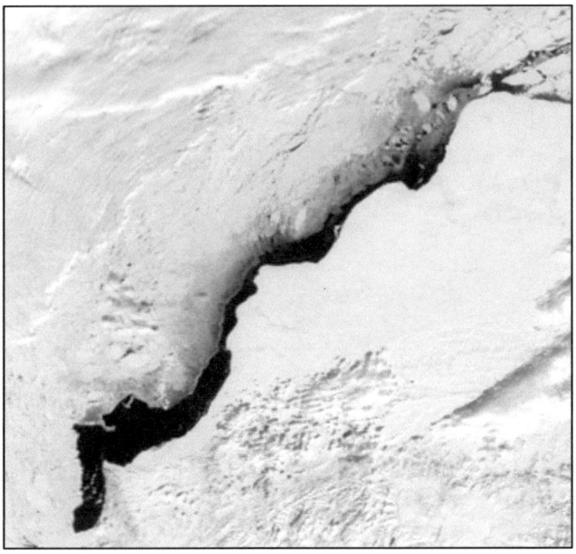

Figure 2.4 Segment of a Moderate Resolution Imaging Spectroradiometer (MODIS) image from March 18, 2002, showing the Chukchi coast from Cape Lisburne (lower left) to Point Barrow (upper right). The broad stretch of open water (dark color) and new ice along the fast ice edge is a characteristic feature of the Chukchi coast during much of the winter in response to prevailing easterly winds. SOURCE: NASA.

By late June, the Chukchi Sea is often close to ice free, while the Beaufort Sea typically remains over 90% ice covered (Eicken et al., 2006; Figure 2.5). Using satellite imagery from 1996-2004, Eicken et al. (2006) determined a mean date of June 4 as the onset of coastal ice breakup, with total clearing being attained on average by June 18, several weeks ahead of the Beaufort coast.

Based on long-term ice chart interpretations, multiyear ice floes in high concentrations (5/10 or more) are rarely found south of Wainwright and very rarely south of Point Lay. Occasionally, old floes have been observed in low concentrations south of Cape Lisburne, but the southern Chukchi Sea is essentially free of old ice throughout the year. Clusters of generally low concentrations of old ice (2/10-3/10 on average) can occur for short periods of time off the northern Chukchi coast from Wainwright to Barrow. Invasions of significant multiyear ice into this coastal area occur approximately two to three times per decade.

Ice charts for the Beaufort Sea and the northern Chukchi coast around Barrow are currently near real time during the summer period. However, there is a need for a consistent level of ice charting for U.S Arctic marine waters. Currently, National Oceanic and Atmospheric Administration (NOAA)

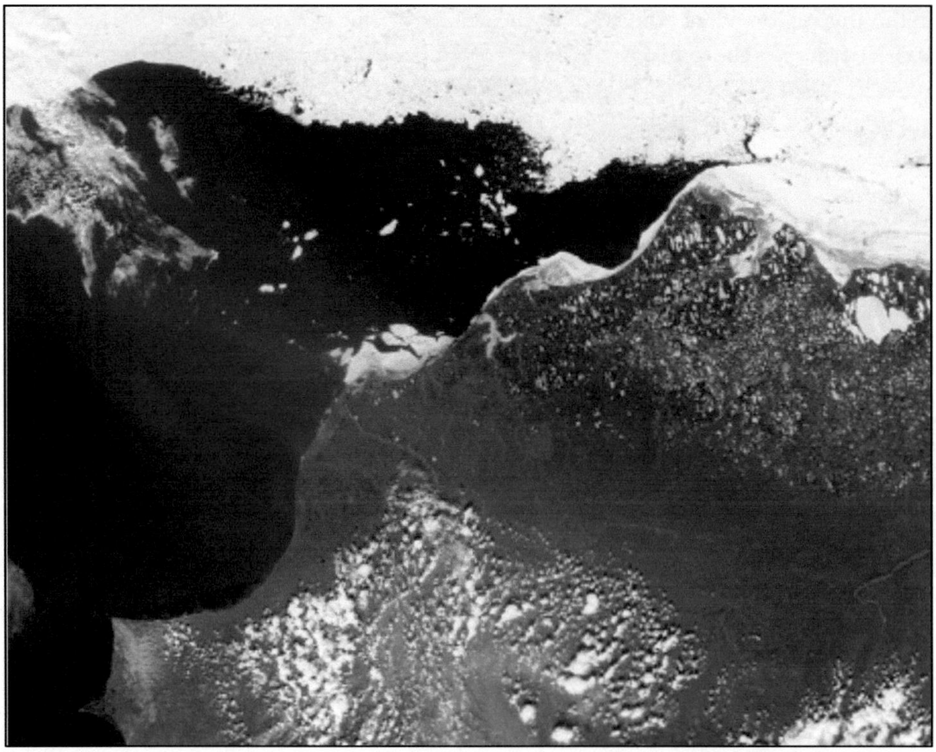

Figure 2.5 MODIS image showing almost complete ice clearing in the Chukchi Sea by June 25, 2005. In contrast, the Beaufort Sea to the east of Barrow (the most northerly point of land in the image) is still choked with very close pack ice. Source: NASA.

Anchorage Ice Desk products and National Ice Center products are not well integrated, with different symbologies and formats. Synthetic aperture radar (further discussed in Chapter 3) could be used for this effort on as frequent a timescale as possible, with higher-resolution coverage in areas of particular significance or interest. A form of accurate ice movement prediction is also needed.

COASTAL PROCESSES AND CHARACTERISTICS

Interactions between terrestrial and marine processes along the Alaskan coast will affect an oil spill's impacts on coastal, nearshore, and offshore environments. This section focuses on a few of these key forces, which include coastline classifications, transport of freshwater into the marine environment, and the role of permafrost with respect to coastline stability and hydrology.

Coastline Classifications

Detailed mapping of coastlines, including geometry and elevation profiles, and knowledge of sediment size, shoreline stability, exposure to wave energy, and vegetation type are critical to understand potential effects of an offshore oil spill and post-oiling recovery of the coastline and associated habitats or protected environments—tundra, barrier islands, beaches and spits, lagoons, lakes, and deltas.

The northern Alaskan coast consists of four main classifications (Hartwell, 1973). Land erosion coasts and wave erosion coasts together comprise approximately 30% of the coastline. Land erosion coasts have bedrock-based, high-relief sea cliffs, while wave erosion coasts have less relief, with cliffs that expose perennially frozen bedrock and ice-rich sediments. The remaining 70% of the coastline is classified as marine or river deposition coasts. Marine deposition coasts resemble wave erosion coasts, except sedimentation processes along the coast have built beaches, barrier islands, and spits. River deposition coasts, by contrast, are built by fluvial processes. About 45% of the coastline is classified as moderate relief (~2-5 m), comprised of cliffs and scarps of wave erosion and marine deposition coasts. Low-relief features (less than ~2 m), such as beaches, river deltas, barrier islands, and spits, make up about 25% of the coastline. High-relief cliffs (~5-8 m) are found along land erosion coasts and wave erosion coasts, while only a few sea cliffs have very high relief features (greater than ~8 m). Together, these comprise about 25% of the coast. The remainder of the coast is open water, such as river mouths and lakes.

Freshwater and Sediment Influx

The annual breakup of Arctic rivers can have great impact on nearshore bathymetry. The rivers draining into the Chukchi and Beaufort Seas are frozen up to nine months of the year, such that almost all of the yearly sediment and freshwater discharge is restricted to short periods in the spring and summer, slightly before and during the spring breakup. In the three-week annual breakup period, the Colville River (the largest river on the North Slope) delivers 43% of its annual discharge and 73% of its total suspended load to the ocean (Arnborg et al., 1967), leading to large areas of flooded land-fast ice. Ice chunks and river runoff erode nearshore bluffs and tundra, and eventual drainage of the floodwaters through cracks in the ice can create significant seabed erosion. In contrast during the winter, no significant freshwater discharge occurs from the Colville River, and seawater encroaches at least 50 km upstream in the delta (Arnborg et al., 1966).

The great seasonality of freshwater and suspended sediment influx could affect oil movement and entrainment in nearshore and offshore environments. Significant amounts of suspended sediments can be deposited on top of nearshore sea ice during flood events. In the case of an oil spill, these sediments could become contaminated and incorporated into the ice, and later redeposited as the ice breaks up and moves. The introduction of freshwater can also affect ocean currents through changes in stratification. Cross-shore salinity fronts established by river runoff can become unstable, causing energetic cross-shelf flows capable of carrying pollutants far offshore (Weingartner et al., 2009).

Permafrost and Coastal Stability

Like many Arctic coastlines, the North Slope is characterized by a continuous layer of permafrost below an active layer, the top soil layer that freezes and thaws over an annual cycle. Permafrost, a perennially frozen layer of ground material that remains at or below 0°C (32°F) for at least two years, often includes ground ice (e.g, ice lenses, layers, and wedges) that forms when water freezes along edges or cracks (UNEP, 2012). Degradation of this ice-bonded permafrost contributes to the high erosion rates observed along Arctic coastlines (Jones et al., 2008, 2009a,b; Lantuit et al., 2011; Romanovsky et al., 2013).

As measured in deep boreholes, permafrost temperatures may have increased by as much as 2°C to 4°C in the early to mid-20th century (Osterkamp, 2007), and by up to an additional 3°C in the 1980s and 1990s alone (Jorgenson et al., 2010). Some studies report relatively stable permafrost temperatures at the turn of the century, but warming trends resumed after 2007 (Romanovsky et al., 2012). Record high warming was measured at most Alaskan permafrost observatories in 2011 and/or 2012 (Romanovsky et al., 2011, 2012, 2013).

These warmer permafrost temperatures increase summer thaw and cause the melting of shallow ice wedges, which decreases the mechanical strength of the coastline sediment and causes the ground surface to subside and form depressions. The result is a lower coastal elevation and a terrain referred to as thermokarst (Jorgenson et al., 2010; Romanovsky et al., 2013). These changes, combined with increased wave energy related to increased areas of seasonally ice-free coastal water, elevated sea surface temperatures, and rising sea levels, have resulted in high rates of coastal erosion and greater inundation of low-lying coastal areas by seawater (Jones et al., 2009a,b; Lantuit et al., 2011). Coastal bluffs along Arctic shorelines are exposed to wave energy that carves out niches at the base of frozen bluffs and eventually causes large blocks of the bluff to collapse (Aré, 1988; Kobayashi et al., 1999; Jones et al., 2009a,b; Lantuit et al., 2011).

The U.S. Beaufort shoreline is underlain with continuous permafrost that is estimated to extend out to at least the 20-m isobath (Brothers et al., 2012); it has also been subject to some of the most dramatic erosion in the Arctic (Lantuit et al., 2011). The ice-rich bluffs have been severely impacted by the cycle of thermal and mechanical erosion described above (Jones et al., 2009b).

Between 1984 and 2011, measurements from Deadhorse, Alaska, at a depth of 20 m document a temperature increase of 2.5°C (Romanovsky et al., 2013). Average erosion rates vary from site to site, with higher erosion rates being more typical along western stretches of the Beaufort Sea and lower rates being reported farther east (Jones et al., 2008, 2009b). Jones et al. (2008) found that a mean erosion rate of 5.6 m/yr between 1955 and 2002, although certain sites had erosion rates as high as 25.9 m/yr. According to a study by Mars and Houseknecht (2007), there are data to suggest that the rate of coastal land loss doubled between 1955 and 2005. Jones et al. (2009b) reported similar findings and stated that in a 60-km stretch of coastline along the Beaufort Sea, the mean erosion rate increased from approximately 6.8 m/yr (1955-1979) to 13.6 m/yr (2002-2007). The impact of permafrost degradation on erosion rates in Alaska along the Chukchi shoreline is less well studied than along the Beaufort shoreline. This may be because earlier studies indicated that erosion rates along the Chukchi coastline have historically been less severe than along the Beaufort (Osterkamp and Harrison, 1982). Furthermore, much of the northeastern Chukchi coast is protected by sandy barrier islands, which largely protect the tundra bluffs from erosion.

Shoreline Mapping

Rapid coastline erosion, thermokarsting, and inundation make it difficult to establish coastal baseline conditions, and shoreline mapping quickly becomes obsolete. Recognizing the importance of a shoreline inventory, more than 20 partner organizations and agencies created a mapping and classification system, ShoreZone,[3] which incorporates low-altitude oblique aerial imagery of Alaska's coastline. ShoreZone has completed mapping of the Beaufort and Chukchi coastline north of the Bering Strait. However, this information needs to be captured efficiently and repeatedly. The use of Light Detection and Ranging (LIDAR, remote sensing that employs a pulsed laser), unmanned aerial vehicles and gliders, satellite stereo pairs, and other new technologies can be used to collect topographic and bathymetric data of the shoreline at high resolutions and can be periodically repeated.

Environmental Sensitivity Index Mapping

NOAA's Environmental Sensitivity Index (ESI) mapping is the accepted standard methodology for classifying shoreline types in terms of their sensitivity to oil spills.[4] ESI maps allow the identification of sensitive areas for oil spill response planning, as well as response action during a spill. The use of ESI maps in spill response can reduce environmental impacts of oil spills and cleanup activities. A number of such maps have been prepared for Alaska, including the North Slope/Chukchi (prepared in 2005) and the Northwest Arctic (2002).

[3] See http://shorezone.org/.

[4] See http://archive.orr.noaa.gov/book_shelf/827_ERD_ESI.pdf.

Hydrographic and Charting Data

NOAA's Office of Coast Survey creates nautical charts for U.S. coastal waters. Arctic shoreline and hydrographic data are mostly obsolete, with limited tide, current, and water level data and very little ability to get accurate positioning and elevation. The nautical charts are of low quality; many were last updated in the 1950s and contain few soundings, little visual navigation, and small-scale, widely spaced surveys (NRC, 2011). Some were based on data last collected in the 1860s, such as the 1:700,000 chart for Kotzebue that was recently replaced by an April 2012 1:50,000 chart of Kotzebue Harbor (presentation by Doug Baird and Jeffrey Ferguson, NOAA, February 2013). There are also issues with tidal and current data (NRC, 2011). The need for more accurate charting in the Arctic was underscored by Presidential Executive Order 13547 (July 19, 2010), which adopted the *Final Recommendations of the Interagency Ocean Policy Task Force,* including the need to address "environmental stewardship needs in the Arctic Ocean and adjacent coastal areas in the face of climate-induced and other environmental changes." Improving navigation and geospatial infrastructure are also goals of NOAA's Arctic Vision and Strategy (NOAA, 2011). In Canada, 2013-2014 priorities for Fisheries and Oceans Canada and other government departments include improving Canadian Hydrographic Services charting in the Canadian Arctic.[5]

As a first step, the Office of Coast Survey released an Arctic Nautical Charting Plan (NOAA, 2013). It engaged stakeholders, state and local governments, and other federal agencies to help determine the needs for future Arctic nautical charting (Figure 2.6). The plans also took into account current vessel traffic patterns, through the use of Automatic Identification System data, and anticipated northern sea routes during ice-free periods of the year. In summer 2012, NOAA began its surveys. Soundings from Dutch Harbor to the most northern portion of the U.S. Arctic coast will be used to prioritize future survey needs. One of the priority areas will be a 1:40,000 chart of the Red Dog Mine terminal on Kotzebue Sound. Survey plans for 2013 included a reconnaissance survey from Kotzebue Sound to Point Barrow (presentation by Doug Baird and Jeffrey Ferguson, NOAA, February 2013).

Satellite imagery and photogrammetry will be used to identify shoreline locations, while multibeam and sidescan sonar data will be collected for water depth and identification of seabed features. The Office of Coast Survey is also using bathymetric LIDAR for nearshore depth data, although it was found to have spotty coverage in areas north of the Bering Sea because very clear water is needed for good results (presentation by Doug Baird and Jeffrey Ferguson, NOAA, February 2013). Two projects in southeastern Alaska and Seward Bay have used a combination of multibeam and bathymetric LIDAR for high-resolution charting[6] and creation of digital elevation models, respectively (Labay and Haeussler, 2008).

Data from sources outside NOAA could help to fill gaps in data collection. This could include bathymetry collected by other federal agencies (NRC, 1994), academia, or industry. While outside data sources can have problems with quality assurance and quality control, analysis of such data

[5] See http://www.dfo-mpo.gc.ca/rpp/2013-14/rpp-op-po-eng.html.

[6] See http://www.fugro-pelagos.com/presentations/Hydro2005_Moyles_Sitka.pdf.

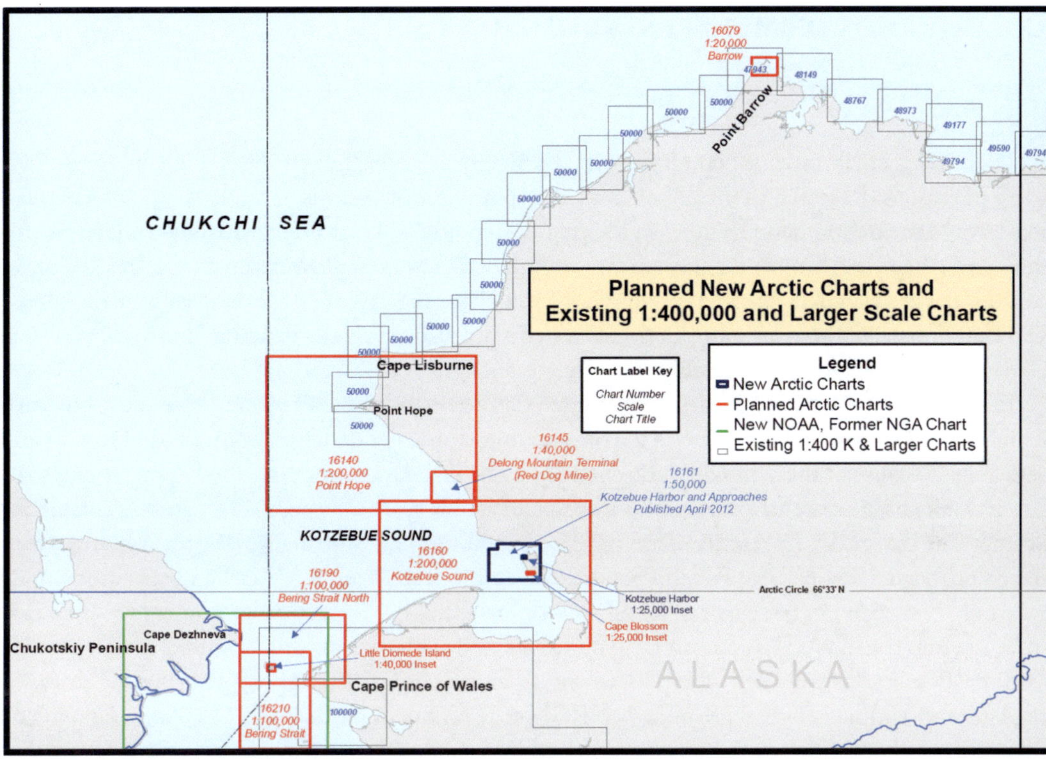

Figure 2.6 Plans for additional nautical charts in the Arctic. SOURCE: Modified from NOAA Office of Coast Survey Arctic Nautical Charting Plan (2013).

could help prioritize regions most in need of updating. NOAA is currently working on guidance to provide to other vessels that might be able to collect data in support of mapping.

Accurate bathymetric charts are part of the infrastructure required for effective oil spill response. The absence of modern charts represents a significant risk to navigation through uncharted obstructions. By extension, shortcomings in nautical charting increase the risk of a vessel-sourced oil spill. Poor charts could also complicate or impede other vessels' abilities to respond to the accident or spill. If a spill was not entirely contained offshore, the ability of large vessels to come close to shore could be compromised. Given the necessity of marine transportation for oil spill response equipment, responders, vessels, and resources, charting infrastructure that provides for their safe and efficient transit is imperative. Poor charting could increase the cost of an oil spill response, as untrustworthy routes or transits require more comprehensive planning. Finally, poor nautical charts hinder preparedness, which could have negative impacts for oil spill response. Several recent reports have recommended that Arctic charting be prioritized (e.g., NRC, 2011).

ECOLOGY AND COMMUNITY STRUCTURE

Baseline Data and Benchmark Needs

Describing environmental conditions in the Beaufort and Chukchi Seas is challenging for a variety of reasons. Research in the region has been sporadic, both temporally and spatially. Although there have been environmental studies in the past, systematic and comprehensive data collection only began with the Outer Continental Shelf Environmental Assessment Program in the late 1970s. In addition, the Arctic environmental regime has been changing rapidly over the past four decades, a trend that is expected to continue. Arctic ecosystems are also especially dynamic due to factors such as seasonal and interannual variability and climate change.

In the event of an oil spill, natural resource damage assessments require "baseline"[7] environmental data that allow federal agencies to compare any injuries or damages from the spill to pre-spill conditions. Because of these rapid Arctic changes, historical data cannot be relied upon to assess the current status of ecosystems or potential impacts of an oil spill, although they can be valuable for documenting trends. In the Arctic, these "baseline" conditions are shifting; yet the term is sometimes used to describe a static state. For this report, the committee chose to describe observational data that would be needed in the event of an oil spill as "benchmarks," or reference points over time that can be measured and used to account for data trends and evolving conditions, rather than as a static baseline. "Baseline" is still used when discussing agencies' activities or reports that use the term.

A recent National Research Council (NRC) study on the impacts of the *Deepwater Horizon* oil spill discusses the difficulty of determining baselines in a dynamic environment, and suggests that ecosystem models that include both biotic and abiotic components could be developed to measure impacts from a stressor, such as an oil spill, in the context of a shifting baseline (NRC, 2013). Development and testing of these models could also assist in the identification of high-priority research and monitoring needs.

Several types of data have been used to approximate baselines for natural resource damage assessments in past oil spills: (1) historical data from either published or unpublished sources (time series are particularly useful for identifying trends and natural variability); (2) environmental data collected at the time of the spill, but prior to exposure to the spilled oil; and (3) environmental data collected from comparable, unexposed sites in the aftermath of the spill. While progress has been made by the Bureau of Ocean Energy Management (BOEM), industry, the North Slope Borough, and the State of Alaska to fill benchmark data needs, there are still gaps in the information. These are described in later sections. The recent increase in activity related to oil and gas in the Arctic

[7] NOAA and the Department of the Interior (DOI) have very similar definitions of "baseline" as it applies to an oil spill or other hazardous spill for a Natural Resources Damage Assessment. For DOI, baseline "means the condition or conditions that would have existed at the assessment area had the discharge of oil or release of the hazardous substance under investigation not occurred" (43 C.F.R. §11.14(e)). NOAA's definition of baseline is "the condition of the natural resources and services that would have existed had the incident not occurred. Baseline data may be estimated using historical data, reference data, control data, or data on incremental changes (e.g., number of dead animals), alone or in combination, as appropriate" (15 C.F.R. §990.30).

offshore has led to a push by federal, state, and local agencies, industry, and academia to characterize benchmark information for Arctic species and ecosystems.

In February 2013, the Interagency Arctic Research Policy Committee (IARPC) produced a five-year Arctic research plan for its 14 federal agency partners, which included in its priorities further study of Arctic marine ecosystems and sea ice (NSTC, 2013). BOEM is currently investing approximately half of its research funding on ecosystem studies in the Chukchi and Beaufort Seas (OSCA, 2013).[8] In addition, the oil and gas industry routinely collects data for exploration, monitoring, and assessment, although much of it is proprietary. In 2008, Shell, Statoil, and ConocoPhillips initiated a multidisciplinary Chukchi Sea Environmental Studies Program.[9] A 2011 agreement between NOAA and those companies created a framework to promote data sharing in the Arctic and led to a release of environmental data from industry to the public through the Alaska Ocean Observing System (AOOS).[10]

Local communities also collect and analyze environmental information through both traditional and Western methods. Through traditional knowledge, communities pass on information regarding changes in benchmark data such as sea ice, winds, and currents; marine mammal, fish, and seabird abundance, distribution, and health; and sensitive information such as haulout locations, good fishing areas, and culturally important sites (Pew Charitable Trust, 2013). For example, the Northwest Arctic Borough has been working on a subsistence mapping project that identifies important ecological areas for subsistence use, including camp sites and species concentration, and collects them in a secure repository (NWAB, 2011).

Synthesis and Monitoring

Although research on Arctic ecosystems has increased in recent years, greater integration and synthesis of the observations is needed to make the information more useful for responding to and assessing damages from an oil spill. Several large synthesis projects are currently under way, including the Synthesis of Arctic Research (SOAR)[11] and Pacific Marine Arctic Regional Synthesis (PacMARS).[12] There have also been recent publications of focused, discipline- or project-specific syntheses (e.g., Bluhm et al., 2010; Hopcroft and Day, 2013). Both SOAR and PacMARS are tasked with integrating the current state of scientific understanding of oceanographic conditions, benthic organisms, lower-trophic prey species (forage fish and zooplankton), seabirds, and marine mammals in the Pacific Arctic, and with providing guidance for scientific research needs in the region. There are a number of database management systems that could be used to integrate and serve these data to interested stakeholders (e.g., the Alaska Ocean Observing System and NOAA's Arctic Environmental Response Management Application [ERMA]). Spies (2011) recommended

[8] See http://www.boem.gov/Environmental-Stewardship/Environmental-Studies/Alaska-Region/Alaska-Studies/index.aspx.

[9] See http://www.chukchiscience.com/.

[10] See http://www.aoos.org.

[11] See http://www.arctic.noaa.gov/soar/.

[12] See http://pacmars.cbl.umces.edu/.

that ongoing species-specific synthesis projects be integrated into a regional ecosystem synthesis, and that monitoring activities across different programs and agencies be coordinated for better efficiency.

Long-term monitoring data provide the best opportunity for tracking ecosystem change and determining benchmarks or baselines for natural resource damage assessments[13] and other purposes. Long-term monitoring needs for the Arctic include habitat use and parameters; abundance, distribution, and health of fish, invertebrates, seabirds, and marine mammals; contaminant concentrations and toxicity evaluations; and subsistence and other human uses (presentation by Mary Baker, NOAA, June 2013). There are different strategies for monitoring, but a few basic tenets apply, especially in areas where sampling can be difficult or expensive. A program cannot monitor everything, everywhere, at all times; instead, a prioritization scheme needs to be determined. One approach for this is to select sentinel species and habitats for regular monitoring. Research on major dynamic processes would inform ecosystem models and help identify essential links in the food web, including prime candidates for sentinel species. Prioritization of future monitoring and research is an essential step for establishing benchmarks. In all instances, it will be essential to incorporate and complement existing monitoring—for example, monitoring of marine mammals for other purposes (such as the Marine Mammal Protection Act).

Monitoring higher-trophic-level species contributes to knowledge about potential changes in lower trophic levels. Changes in prey items (e.g., a shift from benthic to pelagic prey) or in lower trophic productivity (e.g., shifts in species composition or in timing of phytoplankton blooms) can be understood by studying habitat use and food consumption. For mammals, dietary shifts could be monitored through analysis of fatty acids in blubber (Budge et al., 2008; Loseto et al., 2009), stable isotopes in a variety of tissues (e.g., S.H. Lee et al., 2005), scat, or stomach contents from subsistence harvests or stranded animals (Lowry et al., 2004). For example, a 2013 paper analyzing the scat of polar bears revealed prey switching and reflected a change from foraging on sea ice for seal pups to foraging on land for snow geese, goose eggs, and various plant materials (Gormezano and Rockwell, 2013). Monitoring at lower trophic levels can be more streamlined, targeting specific habitats or seasonal trends to optimize monitoring frequency and timing.

Monitoring the diet and distribution of large marine mammals can also indicate the location of important habitats. For example, aerial surveys of gray whales have shown a linkage to areas with high concentrations of benthic amphipods in Barrow Canyon, between Barrow and Wainwright (Moore et al., 2000). Furthermore, concentrations of walruses have been observed in the area of Hanna Shoal, where the benthic community is dominated by gastropods, bivalves, and polychaetes (Schonberg et al., 2014).

Bering Strait and Chukchi and Beaufort Sea Ecosystems

The Chukchi and Beaufort Sea ecosystems have high benthic production in contrast to temperate ocean ecosystems, which are typically dominated by pelagic production. Zooplankton grazing consumes a smaller proportion of phytoplankton (e.g., Campbell et al., 2009; Sherr et al., 2009) with much of the remainder sinking to the seafloor, where it supports a robust benthic community. The

[13] See http://www.darrp.noaa.gov/about/nrda.html.

benthic macrofauna—dominated by polychaetes, bivalves, and amphipods—are a major prey source for seals, walruses, diving birds, and some whales. Fish biomass in the Chukchi Sea is relatively low compared to the Bering or Barents Seas, a reflection of lower zooplankton density, higher export of primary production to the benthos, and possibly the colder water regime (Hunt et al., 2013). Food chains are relatively short, such that disruptions to the lower trophic level quickly cascade to higher-trophic-level species (Grebmeier, 2012).

Several characteristics of these ecosystems make them sensitive to a warmer climate and to earlier melt, breakup, and retreat of sea ice. For instance, the spring phytoplankton bloom is important for zooplankton, including larvae of the resident benthic invertebrates; however, the bloom timing and location depend on sea ice melt and breakup, as well as the seasonal increase in daylight. As the sea ice dissipates, stratification from meltwater keeps nutrients in the surface waters, while sunlight penetrates deeper into the water column. For many species, sea ice is an essential part of the habitat. At the bottom of the food web, ice algae provide a major source of primary production (Gosselin et al., 1997). In the higher trophic levels, walrus and polar bears depend on sea ice as a platform to access rich feeding grounds offshore, and some birds and seals rely on sea ice for nesting areas and rookeries. For these species, a change in timing could disrupt the sequence of trophic interactions and redistribute productivity, potentially shifting some production from benthic to pelagic communities. Warming and earlier retreat of sea ice could also favor the migration of North Pacific species into the Arctic Ocean, which could potentially disrupt existing ecological relationships.

Given the environmental variability of Arctic ecosystems, assessing potential impacts of an event such as an oil spill requires understanding the ecosystem components and how they are intertwined. Food web models can provide one such approach. For example, a preliminary food web mass balance model of the eastern Chukchi Sea illustrates some of the known species relationships in this ecosystem, especially the outsize contribution of the benthic species at the base of the system (Whitehouse, 2013; Figure 2.7).

Lower Trophic Levels

Microbes perform critical biogeochemical functions in the ocean, from primary production and nitrogen fixation to remineralization of organic materials. Microbial photo- and heterotrophs are the dominant biomass of the Arctic Ocean, forming the base of the food web and nutrient cycles. Small eukaryotic phytoplankton dominate, such as picoplanktonic green flagellates and microplanktonic diatoms. Notably, phytoplankton with smaller cell sizes have been replacing those of larger sizes in the past decade (Li et al., 2009). In comparison to temperate and tropical oceans, cyanobacteria are much less abundant (*Synechococcus*) or undetected (*Prochlorococcus*) in the Arctic Ocean (Cottrell and Kirchman, 2009; Vincent and Quesada, 2012).

Heterotrophic bacterial populations found in Arctic seawater perform the essential function of recycling nutrients through decomposition of organic materials. Populations resemble those from temperate waters, with Alphaproteobacteria, Flavobacteria/Bacteroidetes, Gammaproteobacteria, and Verrucomicrobia constituting more than 90% of communities (Comeau et al., 2011; Teske et al., 2011; Ghiglione et al., 2012).

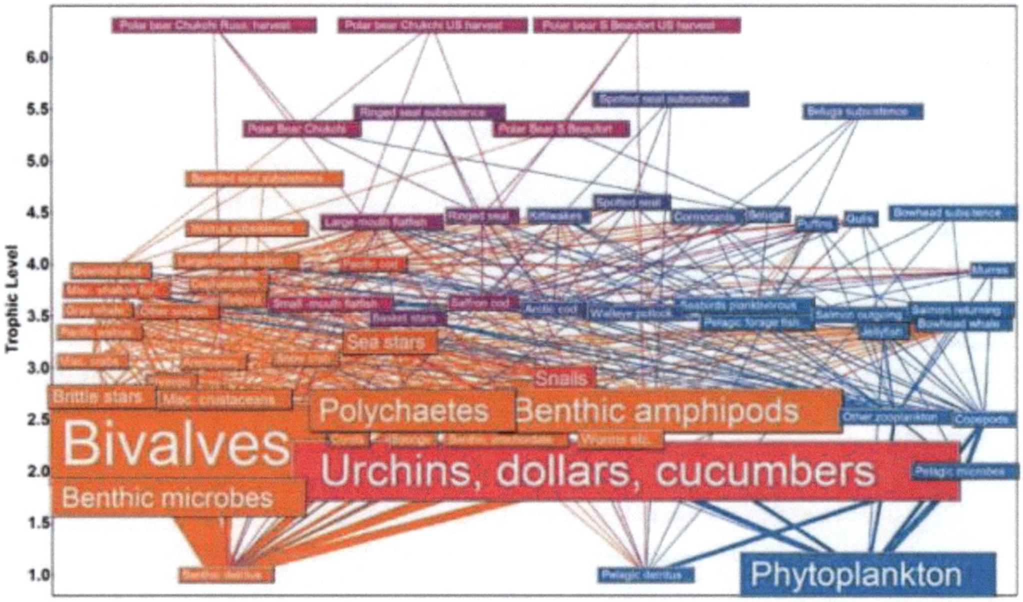

Figure 2.7 Visualization of the eastern Chukchi Sea food web. SOURCE: NOAA Alaska Marine Fisheries Science Center (2012).

Seasonal changes in microbial abundance and distribution have been observed. Prokaryote abundance in seawater decreased fourfold in the Beaufort Sea from summer to winter, but no significant seasonal shift in abundance was observed in the Chukchi Sea (Mrinalini et al., 2012). There was no observed pattern in seasonality or distribution of the major bacterial groups in either region. The community of organisms within brine channels (viruses, bacteria, algae, and small invertebrates) are exposed to fluctuating and physiologically challenging salinity conditions during ice freezing and melting (Gradinger and Zhang, 1997; Faksness et al., 2011).

In addition to the rich microbial community, regions of the Chukchi Sea have some of the highest rates of primary production observed (Arrigo et al., 2012). A continuous influx of nutrients through the Bering Strait, as well as upwelling in Barrow Canyon, sustains high levels of primary production (Sambrotto et al., 1984; Springer et al., 1996; Macdonald et al., 2004). The production and standing stock of phytoplankton changes within and between open water seasons (Kirchman et al., 2009).The first seasonal phytoplankton blooms are associated with sea ice retreat, when the availability of light increases and stratification keeps nutrients in the surface waters (Grebmeier, 2012). Peak phytoplankton growth and reproduction match the timing of sea ice degradation and persist for about two weeks (Cooper et al., 2011). Therefore, earlier sea ice retreat, as seen in recent years, will affect spring phytoplankton blooms (Grebmeier, 2012). The spring blooms provide essential food supporting the early zooplankton growth, but much of the primary production from phytoplankton occurs throughout the ice-free areas during the summer season. In addition, not all primary productivity is dependent on open water blooms. Phytoplankton are transported to the Chukchi from the Bering Sea (Hopcroft et al., 2010), and blooms occur beneath pack ice in the Chukchi Sea because light transmission increases due to thinning of the ice cover and proliferation of melt ponds (Arrigo et al., 2012).

The dramatic blooms of phytoplankton support approximately 350 species of zooplankton in the Arctic, with distributions determined primarily by currents. Oceanic species are dominant in the basin, Arctic species are dominant in the Beaufort Sea, and Pacific species are transported northward through the Bering Strait into the Chukchi Sea. There is a lengthy history of zooplankton sampling in the U.S. Chukchi Sea (Hopcroft et al., 2010), which although sporadic is nevertheless useful for environmental impact assessments and climate research. Zooplankton assemblages are linked to water mass characteristics of temperature and salinity (Hopcroft et al., 2010), with higher abundance and biomass when there is less sea ice (Matsuno et al., 2011). High seasonal and interannual variability in abundance and community structure appears to be normal, a consequence of variability in sea ice retreat and sea surface temperature (Questel et al., 2013). Warmer seawater temperatures in the Arctic could increase grazing rates (Coyle et al., 2007). Reduced sea ice could also allow invasion of Pacific species that might be unable to survive the Arctic winter (Grebmeier, 2012). Increased zooplankton would affect not only the pelagic ecosystem in the Arctic, but also the benthos. Either the unconsumed zooplankton would be deposited on the bottom, enhancing the benthos as unconsumed ice algae currently does, or the enhanced zooplankton would support a shift to a more pelagic community structure than is currently seen in the Chukchi Sea, decreasing export to the benthos and reducing benthic productivity.

Benthic Organisms

There are currently several projects studying benthic macrofaunal invertebrate species (>0.5 mm) in the Beaufort and Chukchi Seas. Both seas were the subject of broad-scale studies in the 1970s. Sampling in the northeastern Chukchi Sea was conducted in the mid-1980s through the present (Grebmeier et al., 1989; Feder et al., 1994; Grebmeier, 2012; Blanchard et al., 2013). There have also been some recent site-specific studies in the Chukchi Sea (Blanchard et al., 2013) and the Beaufort Sea (Dunton et al., 2012). While the broad-scale studies have provided some information on diversity and trends in the region, spatial and temporal variability and limited coverage are not sufficient to create benchmarks to assess impacts from a spill.

In general, macrofaunal assemblages are dominated by deposit-feeding polychaetes and bivalves; in some regions there are dense amphipod beds, such as those in the gray whale feeding area near Barrow Canyon (Blanchard et al., 2013). In addition to larvae from resident populations, the Chukchi Sea receives North Pacific and coastal Alaskan larvae that have been advected north through the Bering Strait.

In the epibenthos, many of the fauna are mobile and can move away from oil pollution related to a spill, but those that are sessile cannot remove themselves from contaminated sediments. Data on epibenthos are collected through trawls, video imagery, and still photos; lack of gear standardization makes comparability an issue across studies. The overall data coverage is low, although there have been several large programs in both the Beaufort Sea and Chukchi Sea since 2000 (Bluhm et al., 2009; Iken et al., 2010; Blanchard et al., 2013), as well as time series in the northern Bering Sea (Grebmeier et al., 2006). For the epibenthic species, it is generally assumed that abundance, composition, and biomass would not change drastically due to an oil spill, unless it occurred during a recruitment phase or a critical part of the reproductive cycle. However, there are still some basic questions regarding epibenthos, including seasonality. While none of the species discussed are commercially harvested, changes to populations due to warming temperatures could lead to impacts on species with subsistence or potential commercial value (e.g., crab, sea urchins, and sea cucumbers). In addition, the epibenthos provides important prey species for walrus and bearded seals.

Fishes

There is only sparse benchmark information regarding northeastern Chukchi Sea ecosystems and Arctic marine fishes (Figure 2.8; Johnson, 1997; Power, 1997; Mecklenburg et al., 2002, 2008). Because the Alaskan Arctic does not contain commercial fisheries (Zeller et al., 2011), there exist no historical commercial fisheries' harvest data from which benchmark data might be reconstructed. Fish-trawl research surveys conducted by NOAA's National Marine Fisheries Service do not occur as far north as the Chukchi Sea. Over the past five decades, fewer than 30 research cruises have collected information on the demersal (bottom-dwelling) fish communities in the eastern Chukchi Sea; of these, only 16 cruises have sampled north of 70°N (Norcross et al., 2013a). Arctic cod (*Boreogadus saida*) was the most abundant demersal (Alverson and Wilimovsky, 1966; Frost and Lowry, 1983; Barber et al., 1997) and pelagic (Eisner et al., 2012) species captured in the western Arctic. The same fish families are found to dominate the northeast Chukchi Sea—cods (Gadidae), sculpins (Cottidae), eelpouts (Zoarcidae), and righteye flounders (Pleuronectidae) (Norcross et al., 2013a). In fish surveys in limited areas of the northeastern Chukchi Sea in 2009 and 2010, Arctic cod was the most abundant species, and the dominant families were cods, sculpins, and pricklebacks (Norcross et al., 2013b), with only small-scale spatial differences shown in fish communities.

Fish in certain coastal areas of the U.S. Arctic have been relatively well studied. A 25-year study on Arctic cisco (*Coregonus autumnalis*), an anadromous whitefish and important subsistence resource, and other species occurred near the Endicott Causeway in the central Beaufort Sea, near Prudhoe Bay (Gallaway et al., 1983, 1997; Fechhelm et al., 1994, 2009). There has been considerable monitoring of Arctic cisco and subsistence fishing in the Colville River Delta (Moulton et al., 2010; von Biela et al., 2011). The North Slope Borough has conducted fish and subsistence surveys in and adjacent to Elson Lagoon.[14] These studies provide a baseline of the presence and relative abundance of nearshore fishes, including whitefish/cisco, salmon, sculpin, capelin, Arctic cod, saffron cod, rainbow smelt, and other species. Subsistence surveys have provided additional information about fish that are important for local communities.

In the Beaufort Sea, few historical data exist for offshore marine fish populations. In 1976 and 1977, bottom-trawl surveys were performed from offshore of Icy Cape northward in the Chukchi Sea into the western Beaufort Sea (Frost and Lowry, 1983). Fish collections in federal waters outside the

[14] See http://www.north-slope.org/departments/wildlife/Fish.php.

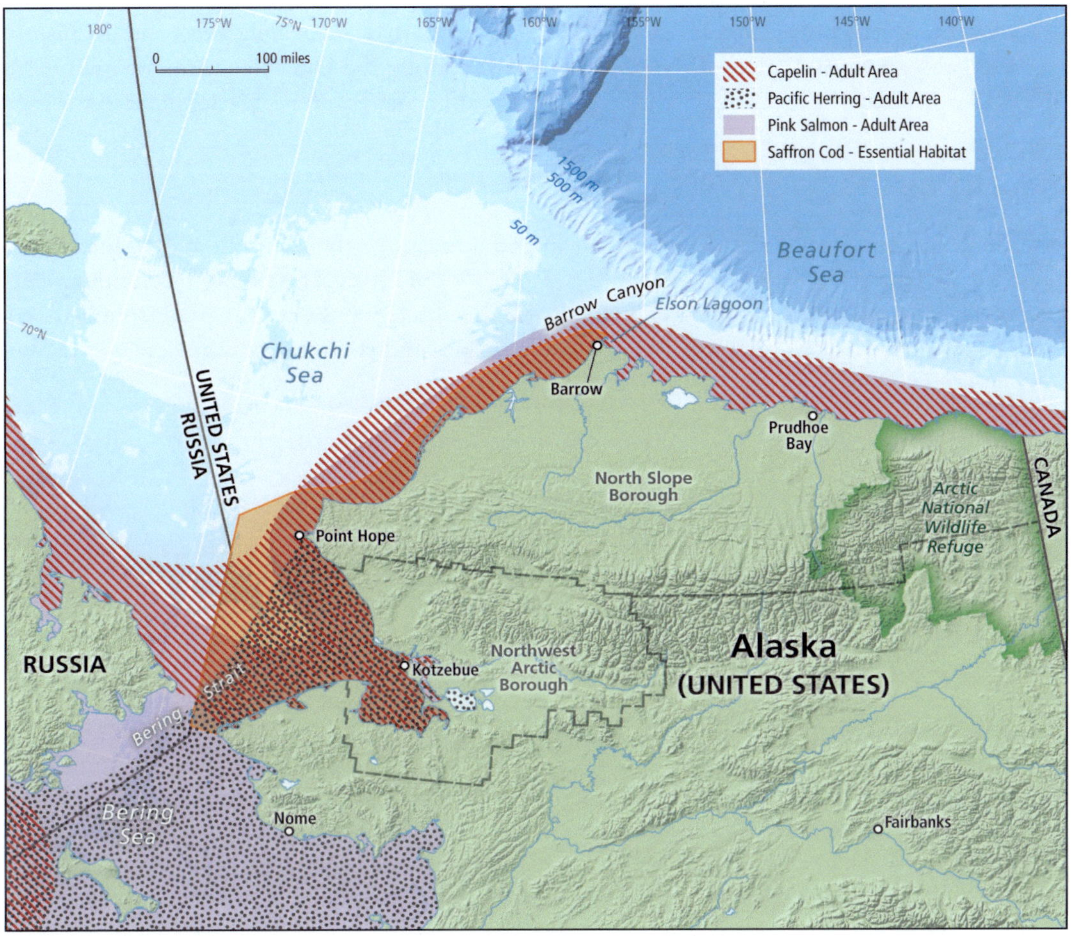

Figure 2.8 Distribution of select fish species in U.S. Arctic waters. Select locations discussed in the text are also shown on the map. Map area corresponds to the red box in Figure 1.1. Data from Arctic ERMA, attributed to Audubon Alaska.

barrier islands consist of a few trawl samples taken in a 1989-1990 Minerals Management Service survey, collected very near the barrier islands and less than 30 km off the coast (Thorsteinson, 1992). However, the Beaufort Sea Outer Continental Shelf Planning Area extends from three nmi seaward across the 100-km-wide shallow shelf. It is unknown whether the limited historical data capture present conditions, especially as the geographic occurrence and trends in subsistence harvests may be changing. Fish assemblages and populations in other marine ecosystems off Alaska have undergone observable shifts in abundance and diversity over the past 20-30 years (Anderson and Piatt, 1999). A 2011 cruise across most of the Beaufort Sea shelf revealed that Arctic cod was the most dominant species of fish. Cods (Gadidae) and sculpins (Cottidae) were among the most abundant families on both the Chukchi and Beaufort Sea shelves (Norcross et al., 2014). A recent BOEM-funded study is assessing new data on fish species composition, distribution, relative abundance, and life history characteristics across 400 miles of the Beaufort Sea straddling the U.S.–Canada border (BOEM, 2013). Camden Bay is a focal area because of oil exploration and potential development in the area.

Marine Birds

Many species and individual birds occur in and adjacent to the Chukchi and Beaufort Seas (Figure 2.9). They include typical seabirds (e.g., loons; procellarids, such as short-tailed shearwaters [*Puffinus tenuirostris*], northern fulmars [*Fulmarus glacialis*]), alcids (e.g., murres, puffins, guillemots), larids (e.g., gulls, terns), shorebirds (including phalaropes and those using nearshore and lagoon

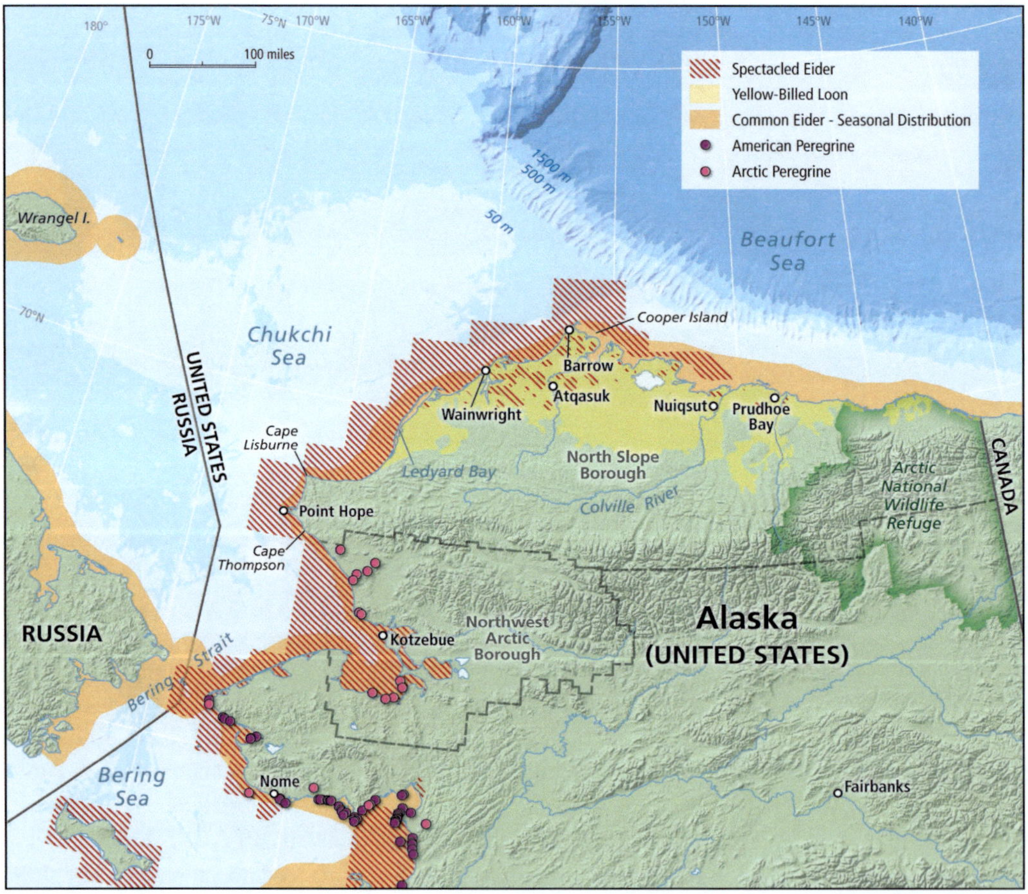

Figure 2.9 Distribution of select bird species found in the U.S. Arctic. Spectacled eider are listed as threatened under the Endangered Species Act. Select locations discussed in the text are also shown on the map. Map area corresponds to the red box in Figure 1.1. Data from Arctic ERMA, attributed to Audubon Alaska (common eider and yellow-billed loon) and NOAA (spectacled eider, American and Arctic peregrines).

habitats), and waterfowl (e.g., eiders, black brant [*Branta bernicla*]). Surveys of these species have occurred through nest colony work, migration counts, aerial surveys, and at-sea surveys.

The largest seabird colonies occur at Cape Lisburne and Cape Thompson. The U.S. Fish and Wildlife Service (USFWS) and U.S. Geological Survey (USGS) have studied murres, kittiwakes, and other birds at those colonies for many years (e.g., Springer and Roseneau, 1978; Springer et al., 1980), although data are limited for monitoring trends of the populations. The primary seabird colony in the Beaufort Sea is at Cooper Island, where black guillemot (*Cepphus grylle*) have been monitored since 1975 (Divoky, 1998) and continue to be observed every summer (Divoky and Harter, 2010). A great deal is known about colony size and trends, demography, nesting success, chick provisioning, behavior, and response to climate change (Divoky, 1998; Harter, 2007; Moline et al., 2008). Other birds nest along the barrier islands, including Arctic (*Sterna paradisaea*) and Aleutian (*S. aleutica*) terns, glaucous gulls (*Larus hyperboreus*), Sabine's gulls (*Xema sabini*), common eiders (*Spectabilis mollissima*), long-tailed ducks (*Clangula hyemalis*), and a variety of shorebirds and songbirds. Some foot surveys for nesting birds have been conducted along the barrier islands (Divoky, 1984; Moitoret and Suydam, 1997), as have aerial surveys in nearshore and lagoon habitats (Johnson, 1993; Johnson et al., 1993; Lysne et al., 2004). Shorebirds have recently been surveyed during premigratory staging in late summer and early autumn along the coasts and deltas of the Beaufort Sea and northeastern Chukchi Sea (Powell et al., 2010).

Adjacent tundra habitats support a diverse bird community. Those birds have been monitored primarily through annual aerial breeding-pair surveys (e.g., Larned et al., 2012; Ritchie et al., 2013), but many site-specific studies have been conducted by federal, state, and local agencies; the oil and gas industry; universities; and non-governmental organizations. The site-specific surveys document population densities and habitat use (e.g., Liebezeit and Zack, 2010; Johnson et al., 2013), breeding biology (e.g., Bentzen et al., 2009; Safine, 2013), and other aspects of the biology of Arctic birds.

Movement and habitat use of larger birds (e.g., loons, eiders, and gulls) have been documented through the use of satellite tagging (Phillips et al., 2006, 2007; Oppel et al., 2009; Phillips and Powell, 2009; Troy, 2010; Schmutz, 2012). Offshore, at-sea surveys for seabirds have included sampling programs between 1975 and 1991 (Divoky, 1987) and 2007 and 2012 (Gall and Day, 2011; Gall et al., 2012). The 2007-2012 program was focused primarily on areas of interest for Chukchi Sea oil and gas exploration. Substantial variation in numbers of individuals and species was seen between the two time periods. These limited data suggest a shift in the seabird community from fish-eating to plankton-eating birds (Gall and Day, 2011; Gall et al., 2012). New approaches to collect observations include the use of high-resolution video surveys from overflights, which are more accurate and repeatable. However, they are less likely to identify or distinguish certain types of birds (presentation by Adrian Gall, ABR, Inc., March 2013).

Audubon and Oceana have identified globally important areas in and around the Chukchi and Beaufort Seas, using colony information, nearshore surveys, at-sea data, and expert opinion (Smith et al., 2012; Christopher Krenz, personal communication, 2013). These areas are important for nesting and foraging for a substantial portion of the global population of a variety of species. Furthermore, Ledyard Bay (northeast of Cape Lisburne) has been designated as critical habitat for threatened spectacled eiders (*Somateria fishceri*) because the area is important for staging, molting, and possibly foraging. Ledyard Bay is also important for king eiders (Oppel et al., 2009), where these two eiders may be foraging on bivalves or other shellfish (Petersen et al., 1998; Suydam, 2000). The North American breeding population of Steller's eiders (*Polysticta stelleri*) is also listed as threatened. They primarily nest between Barrow, Atqasuk, and Wainwright and migrate there from wintering areas in the southeastern Bering Sea. Very little is known about how Steller's eiders use the Beaufort and Chukchi region, although they likely migrate through and stage in nearshore areas.

Population status of most waterfowl and other larger birds is documented through aerial surveys (Larned et al., 2012; Ritchie et al., 2013) and migration counts (Quakenbush et al., 2009). Most species are stable (e.g., Pacific loons, spectacled eiders, long-tailed ducks) or increasing (e.g., greater white-fronted geese [*Anser albifrons*], tundra swans [*Cygnus columbianus*], lesser snow geese [*Chen caerulescens*], black brant [*Branta bernicla*], yellow-billed loons [*Gavia adamsii*]), while a few appear to be declining (e.g., red-throated loons [*Gavia stellata*], mallards [*Anas platyrhynchos*], green-winged teals [*Anas crecca*]). For other birds, such as most shorebirds or songbirds, little is known about population trends, and colony data for most seabirds are not standardized or precise enough to determine trends.

Marine birds in this region are likely to be impacted by climate change, which may lead to a mismatch between the timing of food availability and needs (Moline et al., 2008). These shifts may result in changes to the species that make up bird communities in the Chukchi and Beaufort Seas (Stralberg et al., 2009). Although there is some information on marine birds, numerous data gaps and information needs abound. Key among these is a better understanding of how climate change may impact individual species or populations, as well as the entire bird community. To understand these impacts, or those from an oil spill, key data needs are population trends, distribution, and health and body condition.

Marine Mammals

Considerable research has occurred on marine mammals in the Chukchi and Beaufort Seas, due largely to their importance to the subsistence economies and communities of northern Alaska, but also because many species are listed under the Endangered Species Act. Additionally, they engender a great deal of interest and concern among the general public. The following is an overview of what is known about marine mammals in northern Alaska. Additional details can be found in numerous publications (see BOEM [2013] for ongoing and newly funded studies) and in stock assessment reports regularly updated by the U.S. National Marine Fisheries Service and the USFWS.[15]

There are a variety of marine mammals in northern Alaska. Of the cetaceans, bowhead, gray, and beluga whales and harbor porpoises are commonly found (Figure 2.10). Fin, minke, humpback, and killer whales and narwhals are only occasionally found in the Chukchi and Beaufort Seas; however, sightings of these species seem to be increasing due to changing Arctic conditions. Four species of ice seals (i.e., bearded, ringed, spotted, and ribbon), walruses, and polar bears are found regularly off northern Alaska, while Steller sea lions, northern fur seals, and other pinnipeds are rarely seen (Figure 2.11). Bowhead and beluga whales are well adapted for living in the ice and can be found in the Beaufort and Chukchi Seas from approximately April to October/November. Other cetaceans and most pinnipeds are primarily seen during the open water season, although ringed seals, bearded

[15] See http://www.afsc.noaa.gov/techmemos/nmfs-afsc-245.htm.

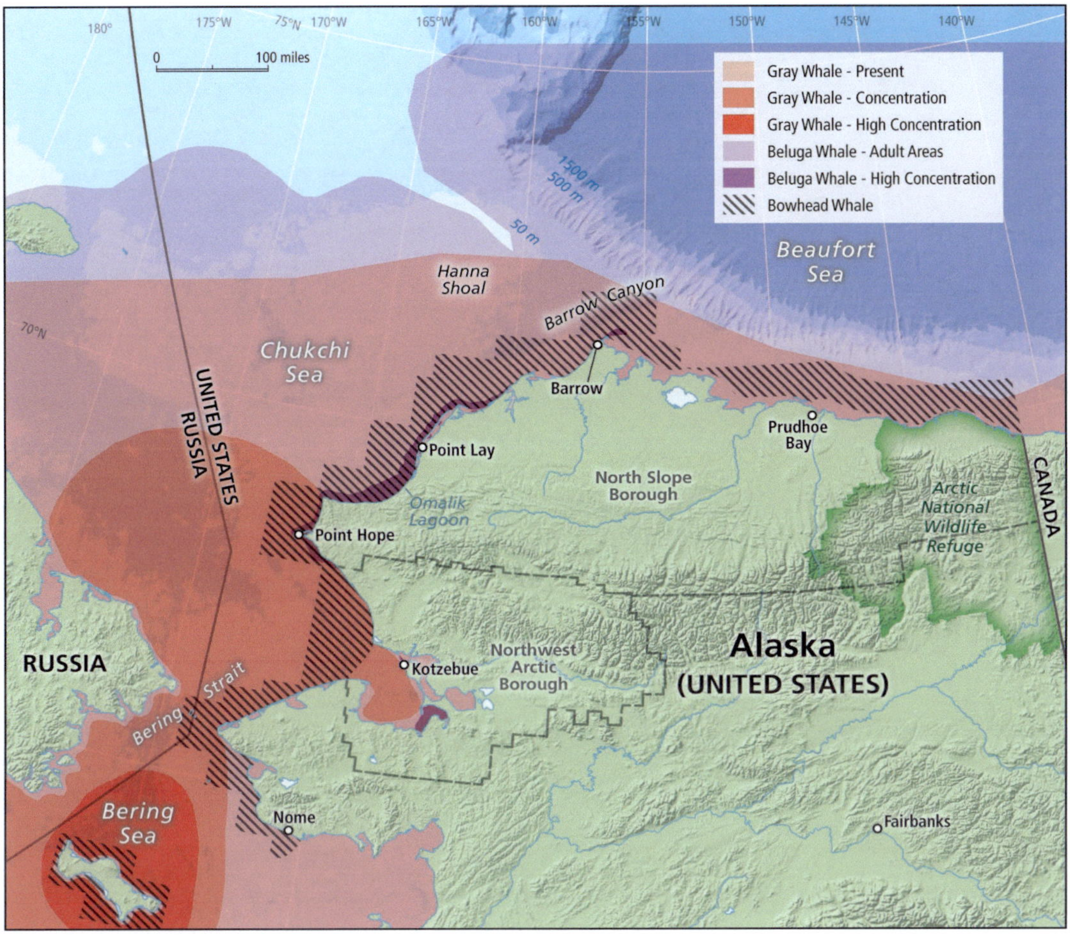

Figure 2.10 Distribution of whales in U.S. Arctic waters. Select locations discussed in the text are also shown on the map. Bowhead whales are listed as endangered under the Endangered Species Act. Map area corresponds to the red box in Figure 1.1. Data from Arctic ERMA, attributed to Audubon Alaska (gray and beluga whales) and NOAA (bowhead whales).

seals, and polar bears can be found when ice is present. Many of these stocks migrate across Arctic waters of the United States, Russia, and Canada.

Population sizes and trends for most U.S. Arctic marine mammals are poorly known, with a few exceptions. The most recent bowhead population estimate (2011) was 16,892, increasing by 3.7% per year (Givens et al., 2012). Based on recent winter migration counts conducted from shore sites in California, the gray whale population ranged from 17,820 in 2007-2008 to 21,210 in 2009-2010 (Durban et al., 2013). Both bowhead and gray whale populations have recovered from or are recovering from overexploitation in the 1800s and early 1900s due to commercial whaling.

The most recent population estimate for the Beaufort Sea stock of belugas was from 1992,

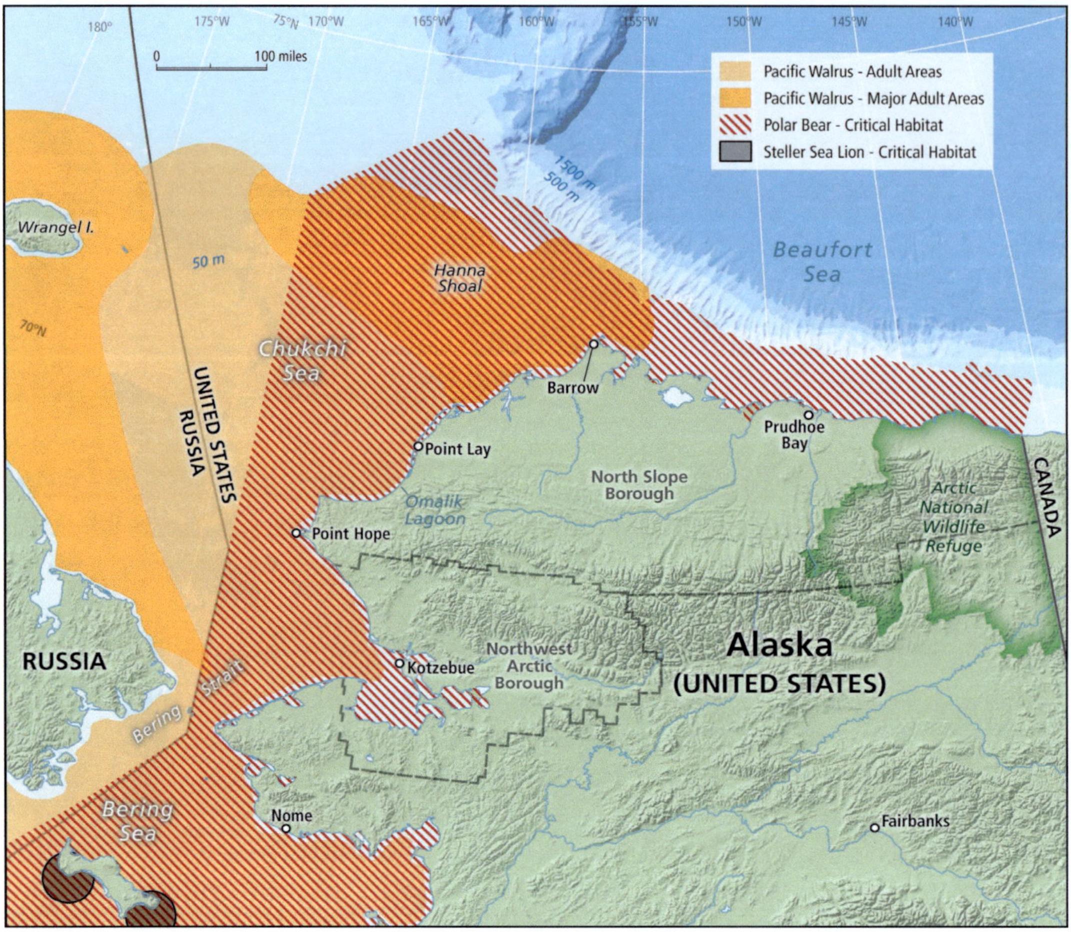

Figure 2.11 Distribution of walrus, sea lions, and polar bears in U.S. Arctic waters. Select locations discussed in the text are also shown on the map. Map area corresponds to the red box in Figure 1.1. Data from Arctic ERMA, attributed to Audubon Alaska (Pacific walrus), NOAA (Steller sea lion), and U.S. Fish and Wildlife Service (polar bear).

when observations from an aerial survey resulted in an estimate of 19,629 (Harwood et al., 1996). When adjusted for availability bias, the estimate increased to 39,258 animals, which may still be underestimated (Allen and Angliss, 2012). More recent surveys suggest that the Beaufort Sea beluga population may be increasing (Harwood and Kingsley, 2013). The population estimate for eastern Chukchi Sea belugas is also outdated and underestimated. Based on surveys from 1989 to 1991, Frost et al. (1993) estimated a minimal population size of 1,200 animals. When corrected for animals not seen, the estimate increased to 3,710 (Allen and Angliss, 2012). Satellite tracking data (Suydam et al., 2001) showed that the aerial surveys for the Chukchi Sea population of belugas only covered a small portion of their range, so it is likely that the population is considerably larger. Data collected in 2012 by the Alaska Beluga Whale Committee and the National Marine Fisheries Service are currently being analyzed and will provide an updated estimate of population size.

The Pacific walrus population was estimated to number about 200,000 animals between 1975 and 1990; however, the estimates were underestimated because they did not adjust for walruses in Russia or under water (Gilbert et al., 1992; Hills and Gilbert, 1994). The results did not provide reliable or consistent data needed for evaluating trends. Using a new approach in 2006, which involved aerial surveys with corrections from thermal imagery and satellite telemetry, the walrus population was estimated at 129,000 (note that the 95% confidence interval = 55,000-507,000) (Speckman et al., 2011). That estimate was also biased low, as the entire walrus habitat was not surveyed. A repeatable survey technique is needed for more precise understanding of the population size and trends of Pacific walruses.

There is no reliable estimate of the Chukchi Sea subpopulation of polar bears, although there have been efforts to develop an aerial survey to obtain a precise and reliable estimate that could be used for monitoring trends (Nielson et al., 2013). The most recent estimate for southern Beaufort Sea polar bears was 1,526 (Regehr et al., 2006). Amstrup (1995) estimated that the Beaufort Sea stock of polar bears increased by 2.4% per year from 1981 to 1992, with a stable population in the 1990s. Based on recent estimates of survival, recruitment, and body size (Regehr et al., 2007), and low growth rates during two years with low ice cover, the Beaufort Sea polar bear population may be declining (Hunter et al., 2007).

There is little understanding of population sizes or trends for other species of marine mammals in the Chukchi and Beaufort Seas (Allen and Angliss, 2012, and references within). While rough population estimates may exist for some species, they are not reliable or precise enough to examine trends, evaluate influences from climate change, or estimate population-level damage in the event of an oil spill.

Habitat use has primarily been documented by aerial and acoustic surveys and satellite tracking. Aerial surveys provide information on the distribution and habitat selection for bowhead, beluga, and gray whales and other marine mammals (Moore, 2000; Moore et al., 2000; Clarke et al., 2013). Acoustic surveys can monitor marine mammals in ways that aerial surveys cannot, because recording instruments can listen for vocalizations in inclement weather, under the ice, and for long periods of time. A great deal can be learned about distribution (Delarue et al., 2011; Clarke et al., 2013), influences of oceanography (Stafford et al., 2013), disturbances (McDonald et al., 2012), and other aspects of the biology of marine mammals. Extensive acoustic monitoring has occurred in recent years in the Chukchi and Beaufort Seas, especially through the efforts of industry (e.g., LGL Alaska Research Associates et al., 2013).

Satellite tracking has provided additional details about the movements of bowhead whales (Quakenbush et al., 2010; Citta et al., 2012), belugas (Richard et al., 2001; Suydam et al., 2001, 2005), spotted seals (Lowry et al., 1998, 2000), walruses (Jay et al., 2012), and polar bears (Garner et al., 1990; Amstrup et al., 2000, 2001). Some progress has been made to tag and track ringed[16] and bearded[17] seals, although additional data are still needed.

[16] See http://www.north-slope.org/departments/wildlife/Ice_Seals.php#RingedSealMovementStudy.

[17] See http://www.north-slope.org/departments/wildlife/Ice_Seals.php#BeardedSeal; see also http://kotzebueira.org/current_projects3.html.

Because marine mammals are harvested for subsistence, much is known about diet, reproduction, age and growth, body condition, and health status. Hunters have allowed scientists to collect biological samples from harvested animals, and information has been collected on the diets of bowheads (Lowry et al., 2004), belugas (Loseto et al., 2009), seals (Lowry et al., 1980; Quakenbush et al., 2011), and walruses (Sheffield and Grebmeier, 2009). Life history traits are known for bowheads (George, 2009) and belugas (Suydam, 2009) and other marine mammals. Data for a variety of contaminants are available for many species (e.g., O'Hara et al., 1999; Woshner et al., 2001a,b, 2002).

There are many concerns about potential impacts from a changing climate—in particular, that rapidly retreating and thinning summer sea ice may have substantial impacts. Some species, such as those that are ice dependent or ice associated, may have adverse population-level reactions to climate changes, while others adapted to subarctic conditions may benefit. Polar bears, ringed seals, and bearded seals, for example, were listed under the Endangered Species Act because of concerns about negative impacts from predicted future impacts to sea ice and snow (including loss of habitat and access to food sources), although other populations may be resilient to changes (Moore and Huntington, 2008). For example, the bowhead population continues to increase despite the dramatic changes in sea ice observed in the past 10-20 years. Other species, such as fin, humpback, and minke whales seem to be increasing in numbers in the Chukchi Sea, probably because of the changing environment. Climate change impacts to marine mammals may occur through habitat changes (Laidre et al., 2008), including less ice and changing prey, changes in timing or distribution of prey populations (Bluhm and Gradinger, 2008), exposure to novel diseases (Burek et al., 2008), and possibly increased predation from killer whales and competition from more southerly populations that are expanding northward (Moore and Huntington, 2008).

Another possible change with decreasing and thinning sea ice could be an increase in commercial shipping through the Arctic (Smith and Stephenson, 2013). Projections of increased vessel traffic could lead to acute or chronic impacts from oil or chemical spills, anthropogenic sound, increased entanglement in marine debris, and ship strikes. Shipping is also associated with the introduction of invasive species, predominantly through the exchange of ballast water.

Even though there is considerable information about marine mammals in U.S. Arctic waters, there are still significant data needs, many of which are directly related to predicting and assessing possible damages from spilled oil. They include population size and trends, especially for seals, walruses, and polar bears; movements, timing, and habitat use, especially for seals; climate change impacts; health status, including exposure to spilled oil and other contaminants; biomarker data related to exposure to spilled oil and other types of stress; and ability to separate possible impacts of climate change versus spilled oil or other contaminants.

Areas of Biological Significance

In the Arctic and elsewhere, areas of biological significance have disproportionally high productivity, biodiversity, and rates of change of pelagic and benthic marine life relative to other surrounding areas (Grebmeier et al., 2010). These areas may include places with high primary or secondary productivity, spawning, nesting or calving areas, rearing areas for young animals, migration routes, or feeding areas. In the winter, sea ice habitats are used by polar bears and ice seals. In some cases, female polar bears may den on the sea ice. Other animals forage for ice seals, which are polar bears' primary prey. Some ice seals migrate south to the Bering Sea during winter, but many ringed and bearded seals remain in the Chukchi and Beaufort Seas to feed, give birth, and molt. Areas near leads or polynyas are likely important for seals and bears throughout the winter.

In spring, lead systems provide habitat for bears and seals as well as migration corridors for marine mammals (especially bowhead whales and belugas) and marine birds. As spring progresses, leads are the first areas for phytoplankton blooms. Recent research has documented incredibly large ice algae blooms under pack ice in the Chukchi Sea (Arrigo et al., 2012). Primary productivity in those areas is among the highest measured in marine systems.

In open water season, nearshore areas are used by a variety of animals. Birds use the habitats for nesting, brood rearing, molting, feeding, and migration stopovers. Marine mammals use specific areas as haulouts for spotted seals and walruses. Belugas can use cobble beaches near Omalik Lagoon, located south of the village of Point Lay, for molting, calving, and feeding. Breaks or passes in the barrier islands, especially in the Chukchi Sea, are also used. Gray whales migrate to the Chukchi Sea in the summer to feed; they can often be found in the shore zone where they appear to rub against the seafloor, possibly in an effort to rid themselves of parasites and barnacles.

There are specific geographic areas associated with some animals. Hanna Shoal is an important foraging area for walruses because of the abundant benthos (Figure 2.11; Clarke et al., 2013; LGL Alaska Research Associates et al., 2013). Oceanographic conditions in and adjacent to Barrow Canyon aggregate fish and invertebrate prey (Rand and Loggerwell, 2010) for belugas and bowheads (Figure 2.10). Ledyard Bay is critical habitat for spectacled eiders (Figure 2.9), as molting birds aggregate and probably forage there in the later summer and early autumn (Petersen et al., 1999). Some king eiders and long-tailed ducks also use Ledyard Bay for molting, and probably also for foraging (Oppel et al., 2009). The inside barrier islands along the Beaufort Sea coast are molting habitat for long-tailed ducks (Lysne et al., 2004) and migration staging areas for red and red-necked phalaropes. Deltas along the Beaufort Sea coast are nesting habitats for geese, particularly lesser snow geese and brant, and yellow-billed loons in some deltas (Ritchie et al., 2000). The Mackenzie and Colville River Deltas, and the nearshore areas between the two, are also important for Arctic cisco, a major focus of the Nuiqsut fish harvest (Figure 2.8; Moulton et al., 2010). The shelf break in the Beaufort Sea is important because belugas aggregate there for feeding (Suydam et al., 2005; Citta et al., 2013). Bowheads may also migrate along the shelf break in the autumn but are more frequently found in shelf areas, influenced by ice concentration (Moore, 2000; Quakenbush et al., 2013). Coastal areas of the Arctic National Wildlife Refuge, which is managed by the USFWS, are important for denning polar bears (Durner et al., 2003).

In the Canadian Beaufort Sea area, the Mackenzie River and its delta are especially important for some marine and anadromous fishes, marine birds, and beluga whales (Harwood et al., 2005). Birds and marine mammals reside here in the open water season, but to a lesser degree in the winter with the exception of ringed seals (Stirling et al., 1982; Smith, 1987). In northeastern Russia, Wrangel Island is a vital denning area for Chukchi Sea polar bears, haulout for walruses, and nesting area for lesser snow geese. The northern coast of Chukotka is also an important haulout area for walruses, and the nearshore waters of northern Chukotka are important for feeding bowheads (Moore et al., 1995), especially in later summer and autumn, and lagoons and bays are used by staging or molting Steller's eiders and other species.

To assess changes in areas of biological significance in the western Arctic between the northern Bering Sea and the eastern Chukchi Sea, an international Distributed Biological Observatory (DBO) was formed. The basis of the DBO is repetitive sampling to gain better understanding of biodiversity, areas with high productivity, and rates of change.[18] Five locations were selected for annual pilot studies between 2010 and 2013, successfully sampling oceanographic conditions, chlorophyll and nutrients, phytoplankton and zooplankton, benthos, and higher trophic levels along a series of transects.

Effects of Oil on Arctic Ecosystems

There are no simple ways to anticipate the effects of an oil spill on an ecosystem. The effects may be acute, as observed in mortality events directly following a spill, and/or chronic, as observed in the population trajectory of species over time or through changes in community composition. Dose-response studies and assessments of sublethal effects on individual species can provide insights into the relative sensitivity of various Arctic species to oil, but they are not always designed to replicate environmental conditions encountered by the organisms after a spill. Much of what has been learned about the acute and chronic effects of oil on ecosystems has come from studies of previous spills or from studies of oil exposure in experimental mesocosms. A summary of the acute and long-term effects on fish, marine mammals, and birds from research on some of the major spills, with particular relevance for high-latitude regions, is available (SL Ross Environmental Research Ltd. et al., 2010). The report illustrates the difficulty in attributing long-term ecosystem or species changes to a spill event, especially in the context of other environmental changes. This could be particularly difficult in the Arctic because of rapid rates of climate change and limited information on the effects of oil on Arctic species and ecosystems. However, effects on organisms could continue in environments where oil persists for years, as was observed for some bird species in oiled areas of Prince William Sound after the *Exxon Valdez* spill.

Lower Trophic Levels

Many bacterial taxa known to degrade petroleum hydrocarbons are found in the Arctic, and recent experiments confirm that fresh Arctic Ocean water contains bacteria that grow on and degrade Arctic Shelf oil at –1°C (McFarlin et al., 2014). In a study of phytoplankton and zooplankton in a freshwater Arctic lake, Miller et al. (1978) found an initial decrease in primary productivity and a suppressed phytoplankton bloom after a controlled experimental spill, but an increase to pre-spill levels afterward. Zooplankton were greatly reduced after the introduction of oil, which may have led to observed changes in the phytoplankton species composition. Other studies have also shown that marine zooplankton are negatively affected by crude oil (Barnett and Kontogiannis, 1975; Berdugo et al., 1977; Miller et al., 1978). Larval stages appear most sensitive to polycyclic aromatic hydro-

[18] See http://www.arctic.noaa.gov/dbo/.

carbons that occur in crude oil, with less impact on later stages, although oil can collect in feeding appendages and bioaccumulation of hydrocarbons can affect egg production and reproductive success (Jensen and Carroll, 2010).

Anthropogenic activities such as oil spills could lead to multiple stressors with direct or indirect impacts on the benthos (e.g., Blanchard and Feder, 2003; Blanchard et al., 2003, 2010). In a highly focused research project that examined the relative impacts of letting an oil slick make landfall on an Arctic beach versus using dispersants in very shallow water (a few meters), filter-feeding benthic organisms rapidly extracted oil from the water column (Humphrey et al., 1987). Deposit feeders were slower to take up oil, but they continued to accumulate oil for as much as two years. Short-term exposure to dispersed oil resulted in temporary accumulation of hydrocarbons in Arctic invertebrates (Mageau et al., 1987). However, this study was deliberately designed to expose benthic organisms to dispersant concentrations far higher than they might experience in a spill in deeper water, where the dispersed hydrocarbon concentrations may be rapidly diluted. Two other studies have documented impacts on benthic organisms in the Arctic due to hydrocarbons (Dauvin, 1982; Blanchard et al., 2011). These studies used small-scale, repeated transects, which are necessary to capture acute effects that may be introduced at the larval stage.

Effects on Marine Birds

Many birds would be vulnerable to spilled oil in the spring lead system, in nearshore areas, or in river deltas during the open water season through direct exposure and contact or through ingestion of oil through preening or consumption of contaminated prey. Because of their importance for resting, molting, feeding, migration, nesting, and/or brood rearing, an oil spill in these areas could have population-level impacts.

Effects on Marine Mammals

Marine mammals that rely on fur for insulation (e.g., polar bears, sea otters) have shown high sensitivity to hydrocarbons, while those that rely on blubber for insulation will be less sensitive to impacts related to thermoregulation. Other health impacts, such as irritation to skin or eyes, ingestion or exposure to oil, or impacts to prey may be similar among species (Geraci and St. Aubin, 1990; NRC, 2003). There are some data available on how oil might affect bowhead whales (NRC, 2003), which could be valuable for assessing environmental changes and possibly for evaluating impacts during or after an oil spill.

It is important to understand other health factors currently impacting some marine mammals—for example, numerous ringed seals and walruses were found sick or dead in northern and western Alaska in 2011. The unknown disease was characterized by skin lesions around the face, flippers, and tail; lethargy; and respiratory distress. An Unusual Mortality Event was declared for ringed seals and walruses but not other species, although some individuals also showed signs of the disease. Fewer sick animals were seen in 2012 or 2013. The cause (or causes) of this disease is still unknown, although investigations continue.[19] Another Unusual Mortality Event was declared for cetaceans in the Gulf of Mexico in February 2010, five weeks prior to the *Deepwater Horizon* oil spill, and lasted through December 2012. The impact of the spill on bottlenose dolphins is still being investigated as part of the Natural Resource Damage Assessment (NRC, 2013) and could provide valuable information for assessing cetacean exposure and health in the future.

NEW AND ONGOING U.S. ARCTIC RESEARCH

In this section, the committee chose to highlight a few new and ongoing research and monitoring efforts that focus on environmental characteristics of the Arctic. Additional research efforts that focus on oil and oil spill response techniques in Arctic conditions are discussed further in Chapter 3.

In July 2012, NOAA's Office of Response and Restoration and several partners initiated an ERMA for the Arctic. ERMA is an online mapping tool that synthesizes a wide variety of data into an interactive map; the aim is to provide assessments and updates of an ongoing situation and improve coordination among responders. Arctic ERMA[20] is a partnership between NOAA (Office of Response and Restoration and Office of Ocean and Coastal Resource Management), the Department of the Interior's Bureau of Safety and Environmental Enforcement (BSEE), the Oil Spill Recovery Institute, and the University of New Hampshire's Coastal Response Research Center. Its aim is to improve access to a wide variety of data from Arctic lease areas. ERMA integrates real-time and static datasets into a single map, where data layers can be turned on and off. The map provides visualization of features including real-time physical conditions, geography and topography, critical wildlife habitats, subsistence use areas, oil and gas wells, other infrastructure, and remote sensing and webcam imagery. With an available Internet connection, ERMA can be used during an emergency to visualize the situation and analyze hazards from the field or from a remote location. Ongoing Arctic ERMA efforts include collaborations to obtain high-priority Canadian datasets, continuing to work with Arctic communities, and integrating additional oceanographic and meteorological data (presentation by Amy Merten, NOAA, February 2013).

A large range of data types and formats exist for the physical and biological Arctic environment, which presents a challenge for effective and timely integration and assimilation. The Advanced Cooperative Arctic Data and Information Service[21] (ACADIS) is a National Science Foundation (NSF)-funded collaboration between NSIDC, the University Corporation for Atmospheric Research, and the National Center for Atmospheric Research that provides data management support for the Arctic community, in particular support for NSF-funded Arctic data collection. ACADIS provides data archival services and is working now to link to other national data centers (including NSIDC remote sensing data) and develop integrated datasets, manage a wide range of data and metadata types, provide better search functionality for data retrieval, and develop protocols for standardized datasets and data visualization tools. All of these elements will improve data assimilation in

[19] See http://www.alaskafisheries.noaa.gov/newsreleases/2011/umedeclaration2011.htm.

[20] See http://response.restoration.noaa.gov/maps-and-spatial-data/environmental-response-management-application-erma/arctic-erma.html.

[21] See http://www.aoncadis.org.

numerical forecast/trajectory models, and support efforts toward real-time integrated visualization of data necessary for responders.

Another NSF-supported initiative, the Arctic Observing Network[22] (AON), is an ocean, atmosphere, ice, and land-based observing system of the Arctic environment. AON includes physical, biological, social, cultural, economic, and system studies of the Arctic. The implementation of AON was in part motivated by recommendations in the 2006 NRC report *Toward an Integrated Arctic Observing Network* to initiate an Arctic Observing Network for the International Polar Year (2007-2008). Data collected under AON are made publicly available as soon as possible and are not subject to any embargo period, allowing improved access to data for a broad range of users. This coordinated effort will be developed further through U.S. involvement in the Sustaining Arctic Observing Networks[23] (SAON) program, an international program of sustained and coordinated pan-Arctic observing and data sharing. Program goals include promoting community-based monitoring efforts and expanding an observing network that can be most effective for a broad range of stakeholders, including communities, policy makers, business, industry, and scientists.

BOEM has launched a large number of ecosystem studies over the past several years in both the Chukchi Sea and the Beaufort Sea, with several more proposed for 2015 (BOEM, 2013). These range from nearshore studies of distribution and habitat use of fish and birds to abundance and distribution of polar bears, bowhead whales, and seals. Much of the research consists of comprehensive marine biological studies with supporting oceanographic components, and many include more than one trophic level. There is an emphasis not only on studies of species identification, abundance, and distribution but also feeding and growth, interaction, and environmental influences. These findings are designed to supply benchmark measurements prior to oil and gas development, and they will also provide critical information to track impacts related to climate change. BOEM has also funded shoreline mapping of the North Slope, as part of the ShoreZone program, to characterize estuarine, intertidal, and nearshore geomorphology (BOEM, 2013).

The Chukchi Sea Offshore Monitoring in Drilling Area[24] (COMIDA) program is one of the comprehensive research programs funded by BOEM. The study is investigating benthic organisms and sediment characteristics, with a new phase that is focusing on Hanna Shoal. As mentioned earlier, there is also a study identifying fish and invertebrates in the U.S. and Canadian Beaufort Seas, which is being done in conjunction with the Department of Fisheries and Oceans Canada. BOEM and NOAA are also co-sponsoring a study of how endangered whales utilize the Chukchi Sea.

Another BOEM-sponsored initiative is investigating the physical properties and currents in the northeastern Chukchi shelf region and exchanges between the Chukchi and Beaufort Seas.[25] The results of this study will be important for evaluating oil spill trajectories and could guide development and validation of oil spill trajectory models. The objectives are to quantify circulation pathways from the shallow Chukchi Sea to the deeper Beaufort Sea; quantify transports through Barrow Canyon; examine the relative importance of prevailing wind forcing, the oceanographic pressure head between the Pacific and Arctic oceans, and the influence of the Earth's rotation in driving the major ocean

[22] See http://www.arcus.org/search/aon.

[23] See http://www.arcticobserving.org.

[24] See http://www.comidacab.org.

[25] See http://dm.sfos.uaf.edu/chukchi-beaufort/background.php.

circulation patterns in the region; understand seasonally changing ocean stratification and how this impacts current systems; and understand processes that transport water between the Chukchi and Beaufort Seas (primarily winds and unstable flow patterns that generate ocean eddies).

The Beaufort/Chukchi Seas Mesoscale Meteorology Modeling Study[26] is also funded by BOEM. Wind forcing, the main driver of surface ocean currents and sea ice motion, and weather patterns in the Beaufort and Chukchi Seas are being monitored, analyzed, and modeled numerically to evaluate and predict oil spill trajectories and assess potential environmental impacts. Local weather systems and winds are complicated in the Beaufort Sea region because of temperature differences between the land and ocean and the steep terrain changes associated with the Brooks Range to the south. Meteorological buoys are being deployed in the Chukchi Sea, atmospheric data are being assimilated into meteorological models, and the mechanisms leading to the formation and dynamics of local weather systems are being investigated.

BOEM's oil spill risk analysis model generates an ensemble of many simulated oil spill trajectories over many years of wind and ocean current data to develop statistics about the risks of an oil spill.[27] Numerical trajectory models estimate trajectories based on velocity data from observations or circulation models; they do not directly assimilate oceanographic or meteorological data, but rely on the degree to which the data represent real-world conditions. For example, as discussed earlier in this chapter, the surface ocean flow over the Chukchi shelf is often not in the direction of the wind because the flow from the Pacific to the Arctic basins is largely driven by the ocean pressure gradient between the two basins. Because surface trajectory models are often forced only by winds or by ocean climatology that is not valid during the time of interest, this key feature of ocean dynamics is not represented. Further significant efforts on model improvement and validation are needed. A BOEM-funded dedicated tracer-release study to examine dispersion in the Chukchi Sea is being planned by researchers at BOEM and the University of Alaska, Fairbanks; this will be interfaced with modeling efforts of NOAA and BSEE.

The Office of Naval Research recently initiated two projects (Emerging Dynamics of the Marginal Ice Zone[28] and Sea State and Boundary Layer Physics of the Emerging Arctic Ocean[29]) to study the retreat of sea ice in the Beaufort Sea and the physical mechanisms in the marginal ice zone. An important part of these studies is wave-ice interaction; wave buoys and ice-mounted instruments are being used to measure ice thickness and atmospheric and oceanic parameters.

In addition to these programs, the European Commission funds basinwide Arctic projects. Current examples are Arctic Climate Change, Economy and Society,[30] which runs from 2011 to 2014, and Ice, Climate, Economics—Arctic Research on Change,[31] which runs from 2014 to 2017. Both have observational and modeling components but also include impacts from oil spills in ice.

[26] See http://mms-meso.gi.alaska.edu/.

[27] See http://www.boem.gov/Environmental-Stewardship/Environmental-Assessment/Oil-Spill-Modeling/Oil-Spill-Occurence-Rate-for-Oil-Spill-Risk-Analysis-%28OSRA%29.aspx.

[28] See http://www.onr.navy.mil/en/Science-Technology/Departments/Code-32/All-Programs/Atmosphere-Research-322/Arctic-Global-Prediction/Marginal-Ice-Zone-DRI.aspx.

[29] See http://www.onr.navy.mil/en/Science-Technology/Departments/Code-32/All-Programs/Atmosphere-Research-322/Arctic-Global-Prediction/Sea-State-DRI.aspx.

[30] See http://www.access-eu.org/.

[31] See http://www.ice-arc.eu/.

OBSERVING NETWORK NEEDS

In an oil spill, early warning is the key to rapid intervention. While efforts to monitor physical processes and ecosystem components are under way (as described in previous sections), there is still a need to bring together disparate data in an integrated fashion. AOOS, AON, and Arctic ERMA have made admirable starts in this effort, but each has a different focus, datasets, stakeholders, and potential or current users. There are also many community-based monitoring programs in Alaska, but they are not well integrated. Existing programs monitor physical conditions (e.g., ocean, ice), species observations (e.g., seabirds, walrus), coastal observations, and water quality and shellfish. Each of these networks has unique data needs and interests in different spatial and timescales. Observing networks in the Arctic are further challenged by very large distances and areas to cover, a lack of in situ instrumentation, and a lack of sensor integration.

An integrated observing network could build on existing networks and efforts, such as AON or AOOS, but could be enhanced with supplemental data from community-based monitoring, in situ instruments, and remote sensing data (discussed further in Chapter 3). To be of value for oil spill response, the network will need to seamlessly interface with marine traffic information systems and monitoring data from oil fields and will need to include comprehensive data on sentinel species such as migratory birds and marine mammals. For local communities, there is need for integrated data networks that can disseminate information needed for subsistence activities, search and rescue, and emergency response. An integrated observing network would also help advance many of the key priorities identified by the Interagency Arctic Research Policy Committee in their 2013-2017 Five-Year Plan (NSTC, 2013).

CHAPTER CONCLUSIONS AND RECOMMENDATIONS

Conclusion: Anticipated Arctic development and its increased potential for oil spills in the region drive the need to improve understanding of current environmental conditions, as well as possible changes to ecosystem parameters in response to a spill. Critical types of benchmark data for oil spill response in the Arctic include:

- Operational near-real-time forecasts and updates of parameters such as winds, waves, ocean currents, ice cover, ice floe size distribution and drift, thickness, as well as derivative products to deliver high-resolution datasets in support of response actions in all weather and seasons;
- Real-time meteorological-oceanographic data from observational platforms (e.g., drilling and production rigs, high-frequency radar, gliders, drifting buoys) to support real-time forecasting and environmental knowledge;
- Identification and monitoring of areas of high biological diversity;
- Identification of rates of change for key species;
- Spatial and temporal distributions and abundances for fish, birds, and marine mammals;
- Subsistence and cultural use of living marine resources;
- Sensitivity of key Arctic species to petroleum hydrocarbons; and
- High-resolution coastal topography and shelf bathymetry.

Additional research and development needs include:

- Meteorological-oceanographic-ice forecast model systems with spatial resolution at 1 km or better and 2- to 3-hour temporal resolution in areas around exploration and production platforms, and
- Integration of traditional knowledge of sea state and ice behavior into the operational forecast environment.

Conclusion: There is a need for a community-based, multiuse observing network in the Arctic that provides long-term, accessible benchmark information (e.g., environmental conditions, local and traditional knowledge, maritime activity) to support oil spill response and other activities. Current Arctic observing networks are fragmented and do not provide a comprehensive regional view. Effort also needs to be devoted to integrating available data from different groups. Such a system could be a collaboration of federal, state, and tribal governments; non-governmental organizations; and maritime and oil/gas industries, and could be organized by IARPC.

Conclusion: The release of proprietary monitoring data associated with exploration activities would further increase knowledge of Arctic environmental conditions and baselines. Making industry and government data more freely available and increasing transparency would bolster the public perception of industry-sponsored research, as would publishing such data in peer-reviewed publications. Where appropriate, communities could also release data that they hold regarding important sites for fishing, hunting, and cultural activities.

Conclusion: High-quality nautical charting is essential for marine traffic purposes and oil spill response in the Arctic. However, shoreline topographic and hydrographic data are mostly obsolete, with limited tide, current, and water level data and very little ability to get accurate positioning and elevation. A lack of high-quality nautical charts from federal agencies can be attributed to inadequate funding and historical prioritization. Data from sources outside NOAA's Office of Coast Survey (e.g., other federal agencies, state agencies, academic research vessels, commercial vessels) could help to fill gaps in data collection. While outside data sources may have quality assurance and quality control issues, analysis of such data could provide information on which areas are most in need of updates.

Recommendation: High-resolution satellite and airborne imagery needs to be coupled with up-to-date high-resolution digital elevation models and updated regularly to capture the dynamic, rapidly changing U.S. Arctic coastline. Nearshore bathymetry and topography should be collected at a scale appropriate for accurate modeling of coastline vulnerability and storm surge sensitivity. Short- and long-term Arctic nautical charting and shoreline mapping that have been identified in NOAA and USGS plans should be adequately resourced, so that mapping efforts can be initiated, continued, and completed in timescales relevant to anticipated changes. To be effective, Arctic mapping priorities should continue

to be developed in consultation with stakeholders and industry and should be implemented systematically rather than through surveys of opportunity.

Conclusion: Ice data and charts are critical needs for marine traffic purposes and for oil spill response efforts. Ice thickness, concentration, and extent need to be integrated into operational use.

Recommendation: A real-time Arctic oceanographic-ice-meteorological forecasting system is needed to account for variations in sea ice coverage and thickness and should include patterns of ice movement, ice type, sea state, ocean stratification and circulation, storm surge, and improved resolution in areas of potential risk. Such a system requires robust, sustainable, and effective acquisition of relevant observational data.

3 Arctic Oil Spill Response Research

Large-scale work on oil spills in sea ice began in the early 1970s in Canada and the United States with the Beaufort Sea Project and the Outer Continental Shelf Environmental Assessment Program, respectively (e.g., Lewis, 1976). In both programs, laboratory and tank research was carried out, but the Beaufort Sea Project also involved the first significant field release of oil under growing sea ice during the winter season. More recently, in 2007-2010 an Oil in Ice Joint Industry Program[1] (JIP) was managed by the Foundation for Scientific and Industrial Research (SINTEF). This large research effort was sponsored by six companies. It included a wide range of laboratory, tank, and field tests, including two cruises in the Norwegian Barents Sea where oil was deliberately released to assess weathering, burning, herding agents, skimmers, and in situ burning (ISB) (Sørstrøm et al., 2010). In 2012, an Arctic Oil Spill Response Technology JIP,[2] with nine participating companies, launched a range of research projects on all aspects of responding to oil spills in the Arctic. This is the largest research program of its kind and is scheduled to continue through 2015 (Mullin, 2012). The U.S. Coast Guard (USCG) recently carried out a series of field trials with icebreakers in the Great Lakes and in the Arctic to test the limits and capabilities of response equipment and remote sensing platforms (Hansen and Lewandowski, 2011; USCG, 2013a,b).

This chapter describes the current state of oil spill response research. It begins with a discussion of a risk-based framework for thinking about oil spill impacts and then provides a summary of the state of knowledge governing our understanding of expected oil behavior in ice. The chapter then moves through a description of various oil spill response options and promising new concepts.

RISK-BASED FRAMEWORK

The committee chose to look at oil spill response in a risk-based framework. A classic risk matrix would depict the risk of an oil spill as a function of its probability (or likelihood) versus the magnitude of its impact (Figure 3.1). Therefore, there are two ways to reduce the risk related to oil

---

[1] See http://www.sintef.no/Projectweb/JIP-Oil-In-Ice/.

[2] See http://www.arcticresponsetechnology.org/.

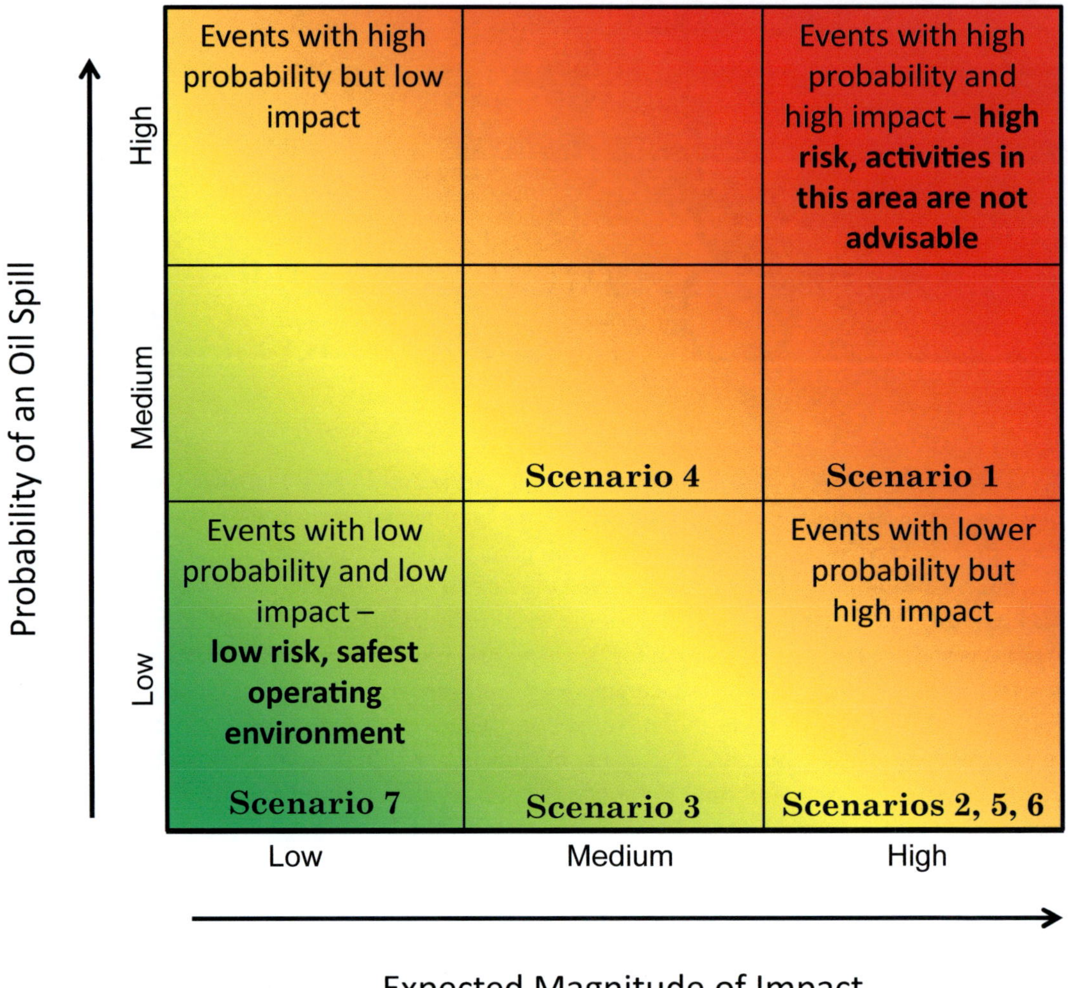

Figure 3.1 Oil spill risk matrix with low-probability/low-impact events in the lower left corner (low risk) and high-probability/high-impact events in the upper right corner (high risk). Risk increases as one departs from the lower left corner of the matrix. Scenarios related to different oil spill response events are mapped onto the risk matrix; numbers correspond to scenarios presented in Chapters 4 and 5. SOURCE: Committee.

spills—reduce the probability of an event, or reduce the magnitude of the impact. Environmental risks include not only low-probability, high-impact events like the *Deepwater Horizon* oil spill, but also small oil spill events with greater likelihoods. The probability of a spill is related to the type and condition of a vessel, pipeline, rig, or storage facility; the accuracy and availability of maps and charts; season, weather conditions, and presence or absence of ice; the behaviors, decisions, and levels of experience of key personnel; and the availability of infrastructure to support spill response (availability of a capping stack, for example). Spill impacts will be related to the amount and type of oil released; meteorological, oceanographic, and geologic conditions, including ice characteristics and cover; the degree of interaction between spilled oil and valued ecosystem elements; and the availability of response infrastructure and trained personnel. The choice and efficacy of oil spill response activities (commonly referred to as "countermeasures") will also affect the magnitude of the impacts.

Risk scenarios were developed to highlight different facets of Arctic oil spill response. Although no formal risk assessments were undertaken, scenarios were developed for discussion purposes and to provide illustrations of the different types of events that could potentially result in an Arctic oil spill response. This is similar to processes followed in a variety of reports (e.g., Arctic Council, 2009, 2013). Scenarios are characterized as having relatively low to high probabilities—based on exposure, frequency, and relative risk levels—and are mapped onto the risk matrix shown in Figure 3.1. Impact levels are characterized as relatively low to high, based on potential oil spill volume or quantity and the ability of resources to reach and respond to the spill. Scenarios are briefly described below, with full discussions found in Chapters 4 and 5.

Scenarios 1 and 4—a passenger cruise or research ship accident and a barge separated from its tow—are determined to be of higher probability due to frequent seasonal operations, often in shallow, nearshore waters. The other scenarios—a large oil tanker accident (Scenario 2), a bulk carrier driven ashore (Scenario 3), a subsea pipeline break (Scenario 5), a well blowout (Scenario 6), and land-based oil tank spills (Scenario 7)—are thought to be less likely events due to existing containment and prevention systems, including established navigation routes for bulk ore carriers. Large tanker accidents, subsea pipeline breaks, and well blowouts are considered to have relatively high impacts, primarily because of large spill volumes and the remoteness of spill locations. The structural failure of a land-based oil tank would be relatively low impact because of the limited spill size, existing spill response equipment, and ability to reach the site with existing onshore resources.

OIL PROPERTIES

Crude oil is composed of a complex mixture of paraffinic, naphthenic, and aromatic hydrocarbons. Oils can differ from each other in a variety of ways, including density and sulfur content. The physical and chemical properties of an oil are not static but can vary between regions, within wells at the same location, and even within a given well over time (EPA, 2011). Key oil properties in cold water environments include measures of the American Petroleum Institute (API) gravity[3] (an indicator of relative density in comparison to water), pour point (the temperature at which a fluid ceases to readily flow), and viscosity. As temperature decreases, viscosity increases and the possibility of going below the pour point becomes more likely. These properties are often considered in early stages of an oil spill response because they usually help define the most effective response options.

There are four standard groupings of oil types (ITOPF, 2013/2014). Group I oils, which include diesel fuel, are nonpersistent—they dissipate rapidly through evaporation and natural dispersion within a few hours and are unlikely to form emulsions, in which water droplets become entrained

[3] API gravity is measured in degrees, and is calculated using the following equation: API gravity = (141.5/SG) − 131.5, where SG is the specific gravity of the petroleum liquid at 60°F.

in the oil through mixing. Group II and III oils will partially dissipate, losing up to 40% of their volume through evaporation. These oils are likely to increase in volume because of their tendency to form viscous water-in-oil emulsions. This also leads to a lack of natural dispersion, especially in Group III oils. Group IV oils have low volatility and are highly viscous. They are highly persistent and are unlikely to evaporate or disperse (ITOPF, 2013/2014). The properties of a fresh oil may change with time, as the petroleum reservoir changes during production. Because of this, a given set of measurements to characterize a fresh oil represents a snapshot in time that may need to be updated. Nevertheless, the classification of oils into specific groups allows broad understanding of how they will behave under different environmental circumstances.

The development of biofuels has somewhat complicated this scheme, as they represent a class of materials that does not readily fit into the categories developed for petroleum products. While the viscosity and relative density of biofuels may be similar to crude oils and petroleum-based fuels, other properties may be quite different, especially as they relate to effectiveness of oil spill response methods. Ethanol from plant sources such as corn, sugar beets, or sugar cane is quite different from crude oil-based fuels because of its infinite solubility in water. In the event of a spill, ethanol would be more akin to a chemical spill rather than an oil spill. Another common form of biofuel is biodiesel, which may be made from either plant- or animal-based materials—animal fats such as tallow and lard; plant oils such as corn, canola, sunflower, and rapeseed; and recycled grease and used cooking oils. Recently, the U.S. Air Force and Purdue University have focused on biofuel for jet aircraft use derived from the *Camelina* plant species.[4]

Biodiesel generally has higher viscosity, flash point, and pour point compared to petroleum-derived diesel, with similar specific gravity (NREL, 2009). Unlike conventional diesel, biodiesel may be suitable for mechanical collection in the event of a spill because of its higher flash point and pour point, especially in colder environments. However, its response to dispersants may be quite different from that of crude oil-derived products because the biodiesel's range of molecular components is narrower.

North Slope Oil Properties

Oil properties and characteristics for specific fields are provided to regulatory agencies by industry when applying for exploration drilling and development permits. For exploration drilling permits, oil properties are estimated from other oils that have been discovered in the region (see Figure 1.2 for locations of many of the fields mentioned in this section). Light (high API gravity or low density) crude oils with 32-57° API gravity and very low sulfur content (0.0-0.2%) have been encountered in Chukchi Sea wells (Burger, Popcorn, and Klondike). These petroleum hydrocarbons are similar to oils found at the Umiat, Alpine, Tarn, and related fields in northern Alaska. In the Beaufort Sea, several of the oil occurrences in nearshore wells like Liberty are related to the medium-gravity (25-30° API), high-sulfur (1.0-2.0%) Prudhoe Bay and Kuparuk oils of northern Alaska. Farther offshore, oil occurrences in the Beaufort Sea (Northstar, Kuvlum) are similar to the high-API gravity (30-40°),

[4] See http://www.purdue.edu/newsroom/releases/2013/Q3/purdue-jet-to-fly-to-international-air-show-powered-by-biofuel.html.

low-sulfur (0.0-0.5%) oils known from the northernmost Arctic National Wildlife Refuge (described by Lillis et al., 1999). The offshore Hammerhead oil field (25 km north of Point Thomson) features an anomalously heavy crude oil (19° API) that has been altered by bacterial degradation, but is probably also related to the offshore Beaufort oils (Banet, 1994; Curiale, 1995; Lillis et al., 1999).

The Bureau of Ocean Energy Management (BOEM) has also identified 97 suspected natural oil seeps, 80% of which are in the Chukchi Sea, in unpublished data collected by industry in the 1980s and 1990s. These seep data could provide suggestions of where additional sampling or analysis might be targeted. However, the available seep data are too elementary for biomarker fingerprinting of oil types and making correlations to oils encountered in exploratory wells or oil fields (presentation by Kirk Sherwood, BOEM, March 2013).

OIL WEATHERING

While properties described in the above section can characterize a crude oil at a particular point in time, weathering after being spilled can change its overall chemical and physical properties. Weathering involves processes that are typically experienced in an open ocean environment—evaporation, dissolution, dispersion, oxidation, emulsification, biodegradation, and sedimentation (Sørstrøm et al., 2010). In the case of a subsea oil spill, surface weathering processes may not be significant if oil does not reach the surface. Key factors that impact weathering include air and water temperatures; the presence of waves, currents, and wind; exposure to sunlight; the presence of ice or snow; and the presence of natural sediments (Figure 3.2). Changes in an oil due to weathering affect spill response options and oil interactions with organisms and ecosystems.

Oil characteristics will also determine to what extent different weathering processes will be significant. Some processes affect oil quickly (e.g., evaporation in a warm environment), while others may take longer (e.g., biodegradation).[5] Evaporation and dissolution cause the loss of the lightest chemical compounds in oil; these smaller-molecule, more volatile compounds contribute to lower viscosity, lower density, and greater solubility. Their loss from an oil slick can have a significant effect on the bulk characteristics of any remaining floating oil.

Evaporation is the most rapid weathering process. It accounts for the loss of 20-50% of many crude oils, 75% or more of some refined petroleum products, but only 10% or less of residual fuel oils (NRC, 2003, 2005). The evaporative loss of a light oil under three different ice coverage levels (open water, 30% ice coverage, and 90% ice coverage) at various current and wave height conditions with different air temperatures (–15°C to about -5°C) was studied by Brandvik and Faksness (2009). They reported that evaporative loss was estimated to be 30% for open water, 25% for the lighter ice coverage, and 19% for the heavier ice coverage, due to differences in oil film thicknesses.

Another important process is emulsification. As oil resides on the water's surface, there is a general tendency for it to incorporate water and form an oil-in-water emulsion. The addition of mixing energy from waves can accelerate the process. As emulsification occurs, there is an increase in volume, viscosity, and water content, each of which can influence the efficiency of response options. For example, ISB loses its applicability once the water content of the emulsion begins to exceed

[5] ITOPF online, 2012; see http://www.itopf.com/marine-spills/fate/weathering-process/.

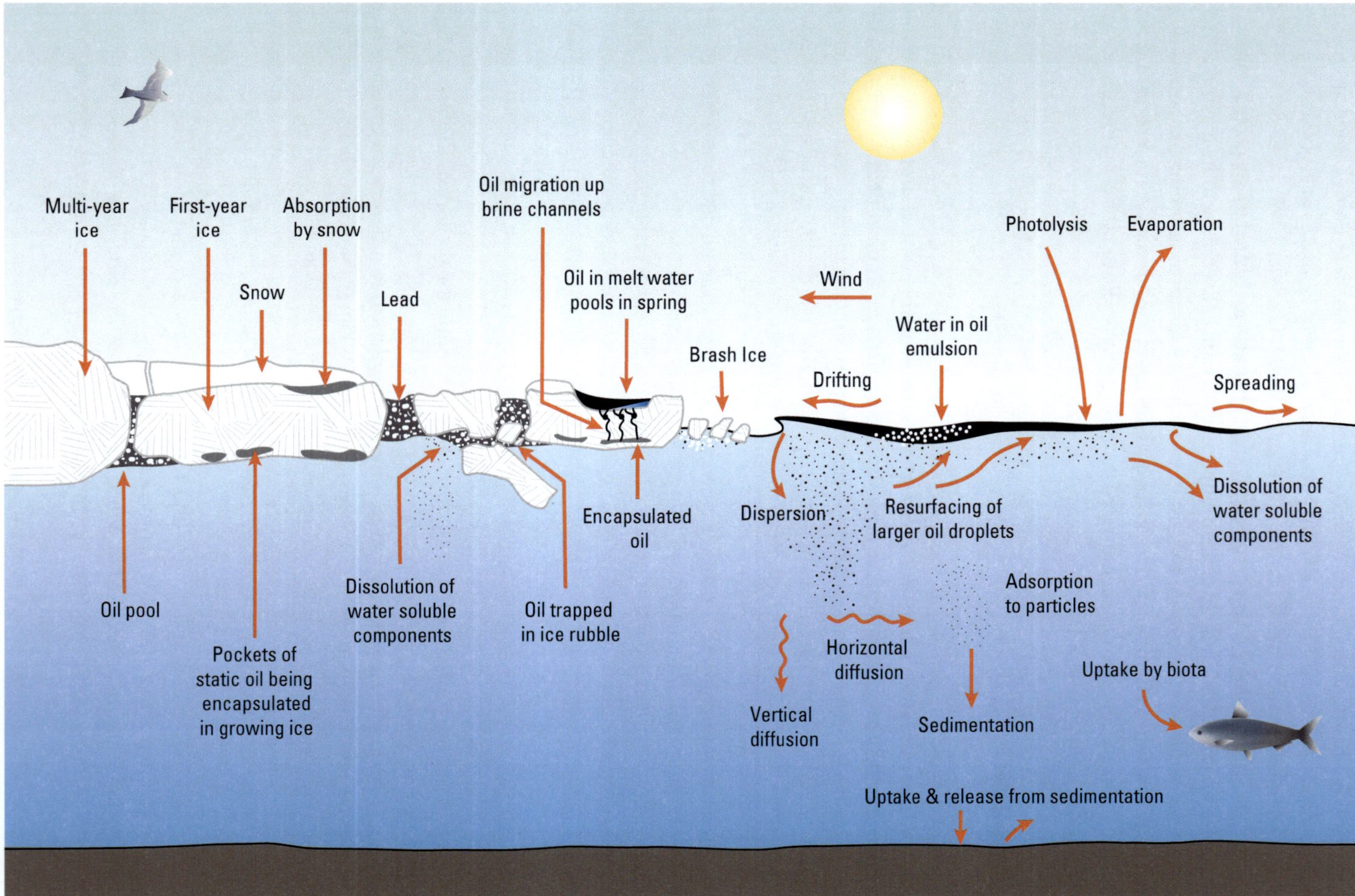

Figure 3.2 Environmental processes that affect oil behavior and weathering in open water and in ice. SOURCE: Modified from Daling et al. (1990) and A. Allen.

30-50%. Additionally, the use of dispersants may become less effective once the emulsion viscosity exceeds a threshold, although this will differ from one crude oil to another (Daling et al., 1990; Federal Interagency Solutions Group, 2010).[6]

Photo-oxidation is an additional process that contributes to the degradation and transformation of crude oil compounds after release to the environment (Garrett et al., 1998; Dutta and Harayama, 2000; Prince et al., 2003b). Light intensity near the water's surface will be lower at northern latitudes due to a low angle of incidence, and the region has a wide range of daylight hours based on the season. Due to longer exposure times during the summer months, photo-oxidation may be a much more important process for oil degradation in the Arctic than in more temperate climates (Serova, 1992; Ivanov et al., 2005).

It should also be noted that not all of the chemicals that make up crude oil biodegrade at the same rate. Higher-molecular-weight components, including polycyclic aromatic hydrocarbons, cyclic alkanes, and naphtheno-aromatic hydrocarbons, may persist for some time.

While weathering trends (Table 3.1) provide some generalities, each real-world case will be different. Weathering may alter the potential toxicity of crude oil components through processes such as evaporation and dissolution of low-molecular-weight components, while photo-oxidation near the water surface could create products not initially present in the oil. There are a number of modeling tools available to assist in weathering predictions,[7] although the underlying assumptions and potential limitations of models and the information from which they operate need to be understood and questioned when necessary. In general, evaporation rates take into account compositional details of specific crude oils. In the event that a crude oil released into the environment is not known completely, the use of another with similar characteristics may be used to gain an approximation of its behavior.[8]

OIL BEHAVIOR IN ICE

The presence of sea ice, discussed in Chapter 2, affects oil weathering processes and the overall behavior of oil in Arctic waters. In many cases, processes that affect traditional oil behavior in open water like evaporation, emulsification, and natural dispersion are slowed down or eliminated for extended periods of time. Laboratory, basin, and field experiments on oil behavior and weathering under Arctic conditions have been conducted independently in Canada and Norway (Sørstrøm et al., 1994, 2010; Brandvik and Faksness, 2009; Buist et al., 2009). Dickins (2011) summarized the behavior of oil in ice, derived from these findings and from direct observations from large-scale field trials dating back to 1972 (McMinn, 1972; NORCOR Engineering & Research Ltd., 1975;

[6] Laboratory assessment of emulsions that were formed during the Macondo Well release indicated that they were susceptible to dispersion at reasonably low dispersant-to-oil ratios. See, for example, work done by SINTEF and documented within (Federal Interagency Solutions Group, 2010).

[7] For example, see the following models: ADIOS2™, developed by the National Oceanic and Atmospheric Administration (NOAA) (Jones, 1997); the Type A model developed by ASA (Reed et al., 1989; French et al., 1996); the OWM model developed by SINTEF (Reed, Singsaas et al., 2001); the OSCAR model developed by SINTEF (1995, 1999).

[8] Environment Canada developed an online database of crude oil properties that is still useful when attempting to determine weathering characteristics: http://www.etc-cte.ec.gc.ca/databases/Oilproperties/oil_prop_e.html.

Table 3.1 Chemical and Physical Changes to Crude Oil from Weathering

Property	Potential Change upon Weathering	Cause
Oil viscosity	Increase (by an order of magnitude)	Loss of low-MW components from evaporation or dissolution
	Increase (several orders of magnitude)	Formation of water-in-oil emulsions (mousse)
Oil specific gravity	Increase	Loss of low-MW components from evaporation or dissolution
Oil slick volume	Decrease	Loss of low-MW components from evaporation or dissolution; loss of mass due to entrainment of oil droplets due to breaking waves
	Increase (by magnitude factor of 3 to 5)	Formation of emulsions (mousse)
Potential toxicity	Decrease	Loss of low-MW components from evaporation or dissolution
	Increase	Formation of photo-oxidation products near water surface

NOTE: MW = molecular weight.

Dickins and Buist, 1981; Nelson and Allen, 1982; Buist and Dickins, 1987; Sørstrøm et al., 1994, 2010; Brandvik et al., 2006; Dickins et al., 2008). Figure 3.2 shows a schematic of different potential oil-ice interactions.

Weathering

Oil spilled during freeze-up will be affected by evaporation, dissolution, emulsification, and natural dispersion to some degree. Most oil spilled during the freeze-up period will remain on the surface of the ice or will migrate up, where it will be affected by evaporation. The evaporation rate is partially controlled by oil slick thickness—thicker oil slicks will evaporate more slowly than slicks in open water. Cold temperatures reduce evaporation rates, as would snow forming a thin film on or covering the oil. Even when covered with snow, oil on an ice surface will lose approximately the same amount to evaporation as it would on water in more temperate waters (Buist et al., 2009).

Brandvik and Faksness (2009) reported that the formation of water-in-oil emulsions under Arctic conditions at a test site inside the ice pack, far removed from the effects of ocean swell, was significantly reduced. The presence of ice effectively dampens both wind wave activity and swell (depending on distance from the pack edge; e.g., Wadhams et al., 1988), increases the thickness of the oil slick, and reduces its surface area. In contrast, within the marginal ice zone at the edge of the pack, there is enough wave action to grind individual floes together and accelerate emulsification. This process was observed by Buist et al. (2009) and Payne et al. (1991) when they did a tank experiment with crude oil spilled in ice in an induced wave environment.

Movement and Drift Rates

Oil trapped within pack ice tends to move with the ice, which is in turn driven by currents and wind; oil in more open drift ice will be less strictly herded by the ice. Under-ice currents in most Arctic nearshore areas are not strong enough to spread the oil far beyond its initial contact with the ice. A 20 cm/s (~0.5 kn) current is needed to initiate and sustain movement of oil under the surface of the oil (Buist et al., 2009).

Spreading

Spreading behavior is one of the properties that is most different between oil spills in ice compared to open water. Oil spilled in the presence of ice is often naturally contained within a relatively small area, which has mostly positive implications for response and recovery options (Buist and Dickins, 1987). Table 3.2 compares the predicted areas and thicknesses covered by a 1,600 m^3 (10,000 barrels [bbl]) crude oil spill on open water, under ice, and on smooth sea ice with and without snow. The table is intended for illustrative purposes to demonstrate the dramatic difference

Table 3.2 Impact of Ice and Snow on the Spreading Rates of Oil, Using as an Example a 1,600 m^3 (10,000 bbl) Crude Oil Spill

			On Smooth Ice	
	Open Water	Under Solid Midwinter Ice	Ice	Snow
Final average oil thickness (mm)	0.016	40 to 90	3	40
Final area (ha)	10,000	7 to 70	50	4

NOTE: The maximum pool depth under solid midwinter ice is determined by the depth of the under-ice depressions, which become deeper as the ice increases and deforms over winter. The final area is determined by both the volume of under-ice depressions and how they fill with oil.
SOURCE: SL Ross Environmental Research Ltd. et al., 2010.

in contaminated areas between oil on water, under ice, and on ice. Under solid midwinter ice, the maximum pool depth varies based on the depth of under-ice depressions, and the contaminated area is determined by both the volume of those depressions and the ways they fill with oil.

The spread of oil is reduced by ice and snow, with resulting oil slicks that are much thicker than those in open water (Dickins, 2011). In practice, an oil slick on open water will spread to cover areas with different equilibrium thickness after some time has elapsed. The majority of the oil is contained within a relatively small, thick patch, while the rest spreads out as a thin film or sheen over a much larger area. In contrast, the maximum contaminated areas under or on ice are predicted to be hundreds to thousands of times smaller. This can be a critical difference when looking at the potential for wildlife exposure, as thin films on the surface create significant risks to waterfowl.

The degree of natural oil containment provided by close pack ice depends on ice concentration and other variables. Generally, the presence of 6/10 ice concentration leads to a slick that is less than half the area of the same oil volume in open water. However, in open drift ice at concentrations less than 6/10, spreading rates approach those for open water (Buist and Dickins, 1987).

Encapsulation, Migration, and Release

Oil density and turbulence tend to govern how much oil becomes incorporated into growing ice, while the oil viscosity is a factor in how it will break down (Dickins, 2011). Heavier fuel oil particles can remain suspended at depth in slush ice; this was observed in the 1979 *Kurdistan* tanker oil spill between Nova Scotia and Newfoundland (Vandermeulen and Buckley, 1985).

Oil spilled under new ice will likely become encapsulated within 12 to 24 hours (NORCOR Engineering & Research Ltd., 1975; Dickins and Buist, 1999; Brandvik et al., 2006). New ice grows beneath the spilled oil, trapping the oil between layers of ice and isolating it from the ocean. Oil may migrate to the surface under thin new ice (less than ~10 cm thick) but will be trapped as the ice solidifies (Dickins, 2011). While lighter fuel oils may surface quickly, heavy fuel oils may be suspended in slush ice. Oil spilled under ice in late winter is unlikely to become encapsulated due to the slowdown in the rate of ice growth. In the spring, the trapped oil migrates vertically though brine channels in the ice sheet (NORCOR Engineering & Research Ltd., 1975; Brandvik et al., 2006; Dickins, 2011).

In an experiment that was part of the 1974-1975 Beaufort Sea Project, over 80% of spilled oil migrated to the ice surface and floated on melt pools by early summer (NORCOR Engineering & Research Ltd., 1975). The oil appeared as essentially fresh crude. Once the oil was exposed to the atmosphere, evaporation occurred at rates similar to spills in southern locations (NORCOR Engineering & Research Ltd., 1975). Oil that is distributed as fine droplets under the ice may migrate more slowly to the surface, a situation that could occur during a subsea blowout with a large gas volume (Dickins and Buist, 1981). In those cases, oil may not be exposed at the surface until the ice begins to melt (Dickins, 2011). As ice melts, encapsulated or surface oil can end up in the water, forming thin oil films before it is naturally dispersed due to wave action.

There are two areas where the knowledge of oil behavior is limited, due to a lack of field observations. The first is the behavior of oil spilled under multiyear ice. The only field experiment designed to explore this behavior provided limited answers (Comfort and Purves, 1982). The second involves processes governing the interaction of oil with new and developing ice during and following freeze-up. This has been addressed by a tank study that was funded through the European Union (Wilkinson et al., 2014).

Developing knowledge of how to prepare for and respond to the possibility of Arctic oil spills is linked to the ability to safely conduct deliberate field-scale oil releases. A number of successful projects over the past four decades have demonstrated that having the ability to conduct deliberate and controlled releases of oil into the marine environment provides an important opportunity to advance the state of knowledge in all aspects of Arctic spill response. When done carefully, these field releases have few or no discernible negative environmental impacts. Important spinoffs from field trials include the opportunity for training of oil spill responders and public consultation to transfer understanding and knowledge.

Over the past 15 years, it has become much harder to obtain the necessary permits in the United States to conduct deliberate oil releases (Dickins, 2011). In the absence of any clear process to permit such work, scientists and engineers have looked toward other countries such as Norway, which has consistently supported this type of research by permitting spills in ice when clear research needs, methods, and goals, combined with responsible cleanup and monitoring plans, have been established (e.g., Sørstrøm et al., 1994, 2010). However, conditions are not completely equivalent to those encountered in the U.S. Arctic. Carefully planned field releases in areas potentially impacted by exploration, such as Alaska, Greenland, Canada, and Russia, could improve evaluation of new response strategies and understanding of oil-in-ice interactions. Dickins (2011) summarized the benefits of previous experimental field releases for research purposes. While there is broad support among industry and other stakeholders, there have been no successful permits to conduct a deliberate field release in U.S. waters since the early 1990s. There has been some interest from the federal government as well. The Interagency Coordinating Committee on Oil Pollution Research has been in discussion with the Environmental Protection Agency (EPA) and other federal agencies, with an aim to improve the permitting process for deliberate oil releases for experimental purposes (Eric Miller, personal communication, 2013). In 2013, the Bureau of Safety and Environmental Enforcement (BSEE) issued a solicitation for a joint industry project to assess the need for a field release of oil, dispersants, and natural gas in the U.S. outer continental shelf, with the possibility of an actual release. BSEE and EPA are also working together to develop a simulant that would successfully mimic the behavior of oil droplets in water.[9]

OIL SPILL SOURCES AND VOLUMES

Three potential point sources for oil spills in the U.S. Arctic are oil and gas wells and pipelines, ships (large oil tankers, bulk carriers, and fuel barges), and land-based municipal fuel storage tanks. The following section discusses likely spill volumes in each of these cases.

If a blowout were to occur in an exploratory well in the Chukchi or Beaufort Seas, the possible rate of uncontrolled flow would be dependent on a number of variables. These include the reser-

[9] See http://www.bsee.gov/Research-and-Training/Oil-Spill-Response-Research/Projects/Project1029/.

Table 3.3 Beaufort and Chukchi Sea VLOS Volumes, Modeled by BOEM for Purposes of Environmental Assessment

Model	Worst Case Discharge (bbl/day)	Oil Discharge at Day 30 (bbl) (Required by Regulations)	Relief Well Period (days)	Oil Discharge at End of Relief Well Period	Model Oil Gravity (°API)
BOEM Beaufort VLOS	69,271	1,140,655	300	3,922,903	26
BOEM Chukchi VLOS[a]	61,672	1,148,300	46	1,552,400	35

NOTE: The relief well period for the Beaufort Sea assumes that ice conditions prevent access to the well throughout the winter.
[a] BOEM (2011).
SOURCE: Presentation by Kirk Sherwood, BOEM, March 2013.

voir characteristics (e.g., porosity, permeability, pressure, temperature, oil viscosity, gas content, and compressibility), wellbore configuration, and the ambient pressure at the blowout discharge point at the seafloor or at the rig. BOEM models "very large oil spills" (VLOSs) and "worst case discharges" using a finite-difference simulator, incorporating data from exploration wells, onshore and nearshore exploration and production wells, likely well designs, known or estimated properties of potential reservoir oils, and geophysical data from seismic surveys (presentation by Kirk Sherwood, BOEM, March 2013). VLOSs represent extreme (improbable, but geologically possible) scenarios that are intended to support assessments with very high environmental impacts. The models estimate daily discharge rates but also report the cumulative amount of oil that could be discharged over the estimated period of time required to drill a relief well (presentation by Kirk Sherwood, BOEM, March 2013). The results of two VLOS models are presented in Table 3.3. In practice, a successful well-capping operation could halt or significantly slow the flow of oil into the marine environment earlier than the time needed to drill a relief well.

Many types of ships have recently been utilizing the Arctic marine environment, including government vessels and icebreakers, container ships, general cargo ships, bulk carriers, tanker ships, passenger ships, tugs and barges, fishing vessels, and vessels related to oil and gas exploration (Arctic Council, 2009). A record of transits through the Northern Sea Route in 2013[10] indicates that some tanker ships carried over 800,000 bbl of oil as cargo, although smaller ships carried as little as 35,000 bbl of diesel fuel cargo. These numbers illustrate the broad range in volume of potential spills from cargo ships, which does not include the fuel oil they carry aboard. In the U.S. Arctic, doubled-hulled barges that provide fuel resupply for Alaskan villages can carry over 6,000 bbl of oil cargo.[11]

The villages store oil, diesel, and gasoline supplies for home and business heating, aviation fuel, and industrial needs for mining and oil and gas production. Because there are long periods between resupply due to sea and river ice, significant volumes of fuel may be stored in relatively close proximity

[10] See http://www.arctic-lio.com/docs/nsr/transits/Transits_2013_final.pdf.
[11] See http://www.marinelink.com/news/ecofriendly-operate335711.aspx.

to the shoreline. Examples include large storage facilities at the Red Dog Mine's Delong Mountain Terminal and tank farms in the community of Barrow.

ARCTIC OIL SPILL COUNTERMEASURES

Arctic response strategies can leverage the natural behavior of oil in, on, and under ice. For instance, ice can bar the spread of oil, reducing spreading rates and leading to smaller contaminated areas (Sørstrøm et al., 1994, 2010; Potter et al., 2012); due to encapsulation or a lack of weathering, oil remains fresher for a longer time; and ice-covered areas generally have less severe wind and sea conditions. Despite the documented effects of climate change leading to later freeze-ups, greater extent of northerly ice edge retreat, and longer summer open water seasons, the Chukchi and Beaufort Sea coastlines are still buffered from oil spilled offshore by a fringe of fast ice for eight to nine months of the year. However, Arctic conditions impose many challenges for oil spill response—low temperatures and extended periods of darkness in the winter, oil that is encapsulated under ice or trapped in ridges and leads, oil spreading due to sea ice drift and surface currents, reduced effectiveness of conventional containment and recovery systems in measurable ice concentrations, and issues of life and safety of responders.

The following sections review the state of knowledge and recent advances regarding key response countermeasures and tools for oil removal under Arctic conditions: biodegradation (including dispersants), ISB, mechanical containment and recovery, detection and tracking, and oil spill trajectory modeling. It should be noted that the highlighted countermeasures are in addition to the "no response" option of natural recovery, a viable response option for some situations.

Although a number of new research developments are discussed in the following sections, this is not a comprehensive account of all ongoing efforts in the field of oil spill response. An extensive list of recent Arctic oil spill research can be found in *Oil Spills in Arctic Waters* (USARC, 2012). It summarizes not only federal efforts but also research and development efforts by non-profit organizations such as the Oil Spill Recovery Institute, industry such as Alaska Clean Seas and the Joint Industry Programs, and international efforts in Canada and Norway.

Biodegradation and Dispersant Use

Biodegradation by naturally occurring microbial communities is a major process contributing to the eventual removal of oil that enters the marine environment (Leahy and Colwell, 1990; Atlas and Bartha, 1992; Atlas, 1995; Brakstad and Bonaunet, 2006). While their numbers may be low in pristine environments, numerous scientific studies have shown that microbes have the ability to rapidly multiply following exposure to oil (Atlas, 1995). Biodegradation of petroleum hydrocarbons has been linked to a diverse community of microorganisms that have large surface-to-volume ratios and the ability to respond rapidly to changes to environmental conditions (ZoBell, 1973; Atlas, 1984; Leahy and Colwell, 1990).

Microbial responses to oil in marine environments are generally dominated by bacteria rather than archaea (Röling et al., 2004). Although fungi are known to degrade petroleum compounds in some marine settings (Zinjarde and Pant, 2002) and their existence has been reported in high-latitude marine environments (Butinar et al., 2011), there have been no studies to evaluate their potential to degrade petroleum hydrocarbons in Arctic marine waters. Populations of indigenous oil-degrading bacteria are likely to be established in Arctic waters due to the presence of natural seeps (Landes, 1973).

The potential of indigenous microbes to degrade petroleum hydrocarbons in marine waters at low temperatures has been demonstrated in field and laboratory studies (Sveum and Ladousse, 1989; Bragg et al., 1994; Prince et al., 2003a; Brakstad et al., 2008; Lee et al., 2009b). Indigenous bacteria in Arctic waters have demonstrated the capacity to degrade petroleum hydrocarbons even at near-freezing temperatures (Brakstad and Bonaunet, 2006). Although the rates of oil biodegradation were lower than those observed at high temperatures (Margesin et al., 2003; Michaud et al., 2004), substantial levels of oil biodegradation by a consortium of bacteria including *Pseudoalteromonas*, *Pseudomonas*, *Shewanella*, *Marinobacter*, *Psychrobacter*, and *Agreia* were observed in nutrient-enriched Arctic seawater (Deppe et al., 2005; Brakstad and Bonaunet, 2006). However, it is important to note that Arctic surface waters are generally nutrient-depleted in summer. The presence of oil degraders does not appear to be a limiting factor in Arctic waters. Despite low population levels, the resident organisms are capable of rapidly responding to changes in environment and energy source. Newer studies suggest that indigenous bacteria in Arctic waters degrade oil faster than previously thought (Stapleton and Sayler, 2000; Whyte et al., 2002; Heiss-Blanquet et al., 2005; Prince et al., 2013). Recent studies with fresh Chukchi Sea water incubated at -1°C gave a biodegradation half-life of 60 days for Alaskan North Slope crude oil (McFarlin et al., 2014). Field trials on pristine Arctic and Antarctic beaches also reveal the presence of indigenous hydrocarbon-degrading bacteria (Grossman et al., 1999; Delille and Delille, 2000; Powell et al., 2005). The loss of oil from sediments at these sites was attributed to several processes, including physical removal, photo-oxidation, and biodegradation.

While microbial metabolism and motility have been measured in brine channels (Junge et al., 2002, 2003, 2004, 2006; Breezee et al., 2004; Faksness et al., 2011; Mykytczuk et al., 2013), biodegradation of oil in marine ice has not been fully investigated. During a preliminary winter field study with crude oil frozen into fjord ice in Svalbard, the bulk oil stimulated the growth of bacterial biomass and oil-degrading bacteria specifically (Brakstad et al., 2008). Another field study in Svalbard showed no significant degradation of oil in ice at subzero temperatures, but at 0°C, melt pool oil samples that were fertilized with inorganic nutrients had a substantial change in bacterial diversity (Gerdes and Dieckmann, 2005). The combination of low temperature and high salt content in the brine channels requires that microbes be both salt (halo-) and cold (psychro-) tolerant. A limiting factor for sealed brine pockets would be availability of oxygen, which would eventually halt biodegradation. Extremely halophilic or halotolerant microbes able to degrade oil have been reported (Diaz, 2008; Al-Mailem et al., 2010), but not yet in cold environments.

Influence of Environmental Factors

Many research programs have studied the influence of various environmental factors on biodegradation rates of oil spilled in Arctic waters. The effect of temperature on degradation rate in nature is less than one might expect, since naturally occurring bacteria are adapted to the temperature in which they develop (Garrett et al., 2003; Macnaughton et al., 2003; Michaud et al., 2004; Brakstad and Bonaunet, 2006; Venosa and Holder, 2007). Psychrophilic bacteria, for example, may have metabolic rates comparable to those of organisms adapted to moderate temperatures. Recent studies using genomic, genetic, and physiological methods have shown that psychroactive bacteria have a number of molecular adaptations that facilitate their growth at subzero temperatures, including some down to -10°C (Bakermans et al., 2012).

A study using Antarctic water showed that the degree of degradation after 50 days of incubation differed only slightly between three tested temperatures (4°C, 10°C, and 20°C) (Delille et al., 2009). Nutrient (nitrogen and phosphorus) availability was identified in this study as the factor limiting oil biodegradation. In a study with low concentrations (2.5 mg/L; 2.5 parts per million [ppm]) of Alaska North Slope oil in Atlantic seawater, to counter the potential effects of nutrient depletion, it was noted that 80% of the saturates and 2- to 4-ring aromatics were biodegraded after 60 days at 8°C (Prince et al., 2013). There are, however, a number of confounding factors—for example, availability and type of bacteria, oil composition, and nutrient availability.

While the majority of marine bacteria thrive in a salinity range of 25-35 parts per thousand (ZoBell, 1973), species living in transition environments are adapted to salinity fluctuations. Diaz (2008) reported the isolation of a bacterial consortium from a North Sea crude oil sample that was capable of metabolizing hydrocarbons in salinities from 0 to 220 parts per thousand, although greater degradation occurred at lower salinities.

The availability of oxygen is important for more rapid removal of petroleum hydrocarbons (Leahy and Colwell, 1990). While a constant supply of oxygen is maintained at the sea surface by wind and waves, oxygen availability may be a limiting factor in subsurface sediments and within the water column. However, there is evidence of biodegradation of some petroleum hydrocarbons coupled with sulfate reduction in marine anoxic sediments (Lovley et al., 1997).

The composition of oil and its physical properties also influence its potential to be degraded by microbes. Saturates have been shown to have the highest rates of biodegradation, followed by light aromatics, while high-molecular-weight aromatics and polar compounds exhibit lower biodegradation rates (Prince, 2010). The same order is expected in cold Arctic waters. Changes to oil properties as a result of weathering have received considerable attention, as they influence the behavior of the oil and its biodegradation potential, and may render remaining components to be essentially nontoxic.

Compounds of low molecular weight in an oil slick are subject to two competing processes: evaporation and dissolution. In general, evaporation is slower, but still significant, in cold seawater compared to temperate seawater. Evaporation of the lower-molecular-weight components also results in increased viscosity of the residual oil (Faksness and Brandvik, 2008), which could negatively affect the ability of oil to disperse into very small droplets, thereby reducing biodegradation. Compared to evaporation, components dissolved from the oil phase are immediately available for microbes to degrade in the water column. In cold seawater, the dissolution of oil compounds is decreased compared to temperate water (Faksness and Brandvik, 2008). This decrease in solubility has been used as an explanation for the observed recalcitrance of hydrophobic compounds in cold water and Arctic conditions. The nonpolar compounds of the water-soluble fraction of crude oils are generally considered to be rapidly biodegraded in the marine environment (Brakstad and Faksness, 2000) relative to other more persistent components such as those found in that grouping of components known as the "unresolved complex mixture" (Meredith et al., 2000; Han et al., 2008).

Garrett et al. (1998) demonstrated that photo-oxidation can create molecules with longer carbon chains (tending to contribute to stabilization of emulsions), while biodegradation tends to break down hydrocarbons into smaller, simpler units. Studies have also shown that the photo-oxidized compounds may have increased potential for biodegradation (Ni'matuzahroh et al., 1999; Dutta and Harayama, 2000; Maki et al., 2001).

Emulsification potentially reduces the biodegradation by lowering nutrient availability to oil-degrading bacteria (Brakstad et al., 2011; Cook et al., 2011). Furthermore, rapid emulsion formation could diminish the rates of evaporation and dissolution, causing retention of more toxic low-molecular-weight aromatic compounds within the residual oil. With sufficient mixing energy, oil associated with surface slicks breaks up into droplets that can become entrained, dissipated, and diluted in the water column (Delvigne and Sweeney, 1988; Fraser and Wicks, 1995; Lee et al., 2001). This natural dispersion process enhances oil biodegradation rates because droplet formation increases the surface area of the oil available for microbial attack and increases nutrient availability (Prince, 1993).

Oil spill trajectory models currently used to predict the fate of oil spills have not been calibrated for the full range of environmental factors encountered in the Arctic. At present, there are limitations in the usefulness of numerical models to predict oil biodegradation in cold marine environments. Detailed temperature-related biodegradation studies are needed to improve fate models, which often rely on inadequate datasets for cold climate spills.

Biodegradation and associated microbial studies from the *Deepwater Horizon* spill are instructive in that they show rapid growth of petroleum-degrading microbes as the oil disappeared, under in situ conditions of 4°C, nutrients, and oxygen. Oceanospirillales (a bacterial order) was dominant in the oil plume at one month (Hazen et al., 2010). By the second month, *Cycloclasticus* and *Colwellia* genera dominated, with *Colwellia* also dominant in flocs of oil and organic matter (Baelum et al., 2012). *Colwellia* strains isolated from the samples grew rapidly on Macondo oil at 4°C, confirming their capability to degrade oil components. When uncontaminated deep water from the Gulf of Mexico was incubated with Macondo oil, 25% of it was degraded at 4°C within 20 days; 60% was degraded when the dispersant Corexit was added. There was an associated enrichment of Oceanospirillales and *Colwellia*, showing the response of a very small population of indigenous biodegrading bacteria to the new petroleum hydrocarbon food source.

While much of the oil from the *Deepwater Horizon* spill was either recovered or degraded, some would have settled to the sediments. Sediments in the *Deepwater Horizon* spill region show an enrichment of Deltaproteobacteria, which are known to degrade hydrocarbons anaerobically, and metabolic signatures of anaerobic petroleum degradation (Kimes et al., 2013). The authors hypothesized that these organisms may degrade some of the more persistent oil components in the water column, though this remains to be tested. While the Gulf of Mexico has different characteristics than the Arctic Ocean, its cold deep waters and sediments may have similar bacterial populations. Furthermore, both have natural oil seeps that allow long-term selection for oil-degrading bacteria adapted to regional conditions.

A major oil spill in the Arctic could lead to a microbial community response similar to that observed after the *Deepwater Horizon* spill, but it is not known how changes in microbial communities would impact the food web, especially the significant benthic communities.

Bioremediation

Research has been focused on the development of Arctic oil spill bioremediation strategies that accelerate the natural biodegradation process. Considering the logistical needs in mounting an Arctic spill response, bioremediation may be a very effective countermeasure as it does not require contaminated waste transport and disposal, large numbers of personnel to apply, or specialized equipment. In addition, it is considered by many to be a more environmentally friendly technology than some other responses. Most attempts have used biostimulation to overcome nutrient and oxygen limitation and to promote oil dispersion, increasing growth conditions for oil-degrading bacteria (Lee et al., 1993, 1997; Prince, 1993, 2005; Swannell et al., 1996; Lee and de Mora, 1999; Prince and Clark, 2004; Al-Darbi et al., 2005; Prince and Atlas, 2005).

Marine waters often contain low concentrations of nutrients (most significantly, nitrogen and phosphorus) that limit the hydrocarbon biodegradation (Lee et al., 1993; Prince, 1993; Lee and de Mora, 1999). Biostimulation activities to increase natural biodegradation in the marine environment have focused on applying fertilizers to overcome nutrient limitation (Prince, 1993; Bragg et al., 1994; Swannell et al., 1996; Venosa et al., 2002; Prince et al., 2003a). Its effectiveness was shown during cleanup operations following the *Exxon Valdez* oil spill, where the addition of fertilizer was shown to increase biodegradation rates by three to five times (Bragg et al., 1994; Atlas, 1995).

Application protocols need to be tailored to local conditions to account for differences in environmental parameters and oil characteristics at specific contaminated sites (Wrenn et al., 1994, 1997a,b, 2006; Venosa et al., 1996; Boufadel et al., 1999; Lee and de Mora, 1999). Some nutrient additions may inhibit microbial activity (Braddock et al., 1997), while excess nutrients can cause detrimental effects such as eutrophication. Certain microbial populations may require different nutrient ratios for optimal degradation of different hydrocarbons (Smith et al., 1998). A study that showed enhanced oil degradation in samples of ice-covered seawater also raised the prospect of treating oil within brine channels of sea ice (Delille et al., 1998).

Bioremediation efforts have been mainly carried out on shorelines and in wetlands, due to the difficulty of maintaining elevated nutrient concentrations in an open water system (Leahy and Colwell, 1990). Biostimulation trials conducted on Arctic beaches of Spitzbergen Island showed that application of a commercial nutrient formulation increased oil biodegradation in coarse but not fine sediments (Sveum and Ladousse, 1989). A later program expanded the scope of this study by testing the effectiveness of several other remediation methods including bioremediation, physical mixing, and surf washing (Lee et al., 1998a; Guénette et al., 2003; Sergy et al., 2003) to enhance the biodegradation of an intermediate fuel oil on sand and pebble shorelines. Over one year, biodegradation rates were approximately doubled in the oiled sediments that were treated with fertilizers, with no acute toxicity measured (Lee et al., 2003b; Prince et al., 2003a). By increasing sediment permeability, physical mixing also appeared to increase microbial activity (Owens et al., 2003). Commercial and experimental nutrient mixtures also enhanced degradation of an Arabian crude oil stranded on intertidal sandy beaches on the Kerguelen Archipelago in the Antarctic (Pelletier et al., 2004).

Chemical Dispersants and Oil-Mineral Aggregate Formation The amount of oil that naturally disperses into the water column after a spill is dependent on the kind of oil spilled, ambient temperature, and the mixing energy and/or release conditions. For example, most if not all of the oil spilled in the midst of a gale with high sea states may be naturally dispersed, as in the 1993 *Braer* tanker incident (Spaulding et al., 1994). When the mixing energy that is provided by waves and wind overcomes surface tension at the oil-water interface, natural dispersion will occur and will break the oil film into variable-size droplets (Potter et al., 2012). Larger oil droplets resurface and coalesce, while smaller droplets stay suspended in the water column, where they will be diluted through turbulence and subsurface currents. These smaller droplets will eventually be subjected to processes of biodegradation.

Chemical dispersants are designed to reduce surface tension at the oil-water interface, which would allow waves to more easily break up the oil into small droplets (generally less than 100 microns) that are rapidly diluted in the water column, down to concentrations where natural nutrient levels are not rate limiting for microbial degradation of the residual hydrocarbons (Swannell and Daniel, 1999; Prince et al., 2013). Dispersant guidelines for the State of Alaska are contained in Annex F of the Unified Plan (ARRT, 2008). Alaska has no pre-approved dispersant use areas.

The biodegradation of chemical oil spill dispersants and the surfactants used in them has been studied for many years (Liu, 1983; Una and Garcia, 1983; Odokuma and Okpokwasili, 1992; Baumann et al., 1999; Lindstrom and Braddock, 2002; Venosa and Holder, 2007; Garcia et al., 2009). Varadaraj et al. (1995) noted that dispersants themselves can enhance the initial rate of hydrocarbon degradation, as oil-degrading bacteria can grow on the dispersant and then colonize dispersed oil droplets. This activity can cause a delay in oil biodegradation because the bacteria may preferentially utilize the dispersant first (Foght and Westlake, 1982; Bunch et al., 1983; Foght et al., 1983).

Dispersion will, in many cases, rapidly dilute oil concentrations below toxicity threshold limits and will increase the oil-water interface for greater microbial interaction and degradation. While primarily used on surface oil spills, in the *Deepwater Horizon* oil spill dispersant was applied via subsurface injection to the plume of oil released from the wellhead into 4°C waters at 1,500 m depth (Atlas and Hazen, 2011; Kujawinski et al., 2011). There has been considerable debate over the effectiveness of chemical dispersants on crude oil degradation at low seawater temperatures. The main concern is that as temperature decreases, chemical processes slow down and oil viscosity increases, making it more difficult to disperse. Generally, dispersants are effective on nonemulsified oil at freezing temperatures as long as viscosity does not increase significantly. Experiments to test the effectiveness of eight dispersants on South Louisiana crude oil (analogous to that released during *Deepwater Horizon*) at 5°C and 25°C revealed that temperature was less critical than expected (Venosa and Holder, 2013). To overcome the viscosity limits of conventional chemical dispersants in cold environments, recent research has focused on higher-viscosity products that increase contact time with the spilled oil (Nedwed et al., 2011) and those with higher concentrations of active ingredients. Wave-basin tests indicate that these products might be used to treat conventional oils with dispersant-to-oil ratios as low as 1:100 (compared to the currently recommended 1:20) and to disperse high-viscosity oils such as heavy crude and fuel oils (Nedwed, 2010).

Recently, the SINTEF Oil in Ice JIP (Daling et al., 2010; Sørstrøm et al., 2010) evaluated the effectiveness of dispersants under Arctic conditions, including cold air and water temperature, ice, and melting ice and river outflows. The presence of ice can increase the length of time that a dispersant is effective by slowing the rate of oil weathering and emulsification. Although wind-wave action that facilitates dispersion in open water is generally dampened by the presence of ice, the interaction of individual ice floes in response to winds and currents can lead to localized upper water column turbulence for more effective dispersion than would otherwise be possible without the presence of ice under similar wind conditions. Mechanical mixing may be needed to overcome the lack of turbulent mixing energy in scenarios involving significant ice cover—for example, vessel propellers or thrusters can provide artificial mixing energy while adding chemical dispersants to oil, an effect documented in tank tests and at sea (Nedwed et al., 2007; Daling et al., 2010). There are limitations to the surface application of dispersants, as vessels are able to cover only a limited amount of slick in a given time, while aircraft provide high coverage rates but cannot optimize dose rate to slick thickness.

There still remains some controversy over the use of chemical oil dispersants, due to concerns over their potential to induce toxic effects. A 2013 National Research Council (NRC) report summarized a number of recent studies on the toxicity of dispersants and dispersed oil on plankton, fish, and shrimp and determined that "there is some evidence that chemically dispersed oil and some dispersant compounds are toxic to some marine life, especially those in early life stages" (NRC, 2013). Contradictory evidence on the toxicity of chemically dispersed versus physically dispersed oil on marine organisms (NRC, 2013) exists as a result of differences in experimental conditions and methodology and the fact that actual exposure concentrations were not fully quantified in a large fraction of previous toxicity studies (NRC, 2005).

There are understandable concerns about the toxicity of the dispersant itself. However, in U.S. waters, only EPA pre-approved products that pass standardized product efficacy and toxicity tests are considered for use. Key factors determining toxicity for a given species are concentration and length of exposure time. Studies have shown that chemical dispersants can increase the exposure of oil to pelagic marine organisms. For example, Couillard et al. (2005) reported that the presence of dispersant caused a two- to fivefold increase in the concentration of total and high-molecular-weight polycyclic aromatic hydrocarbons (PAHs). However, while chemically dispersed oil may be more biologically available, dispersants themselves are rapidly diluted in the open ocean to less than 1 mg/L (1 ppm) within an hour, well below defined toxicity threshold limits (NRC, 1989; Lee et al., 2013).

Chemical dispersants do more than facilitate the transport of oil from surface oil slicks to the water column. Prince et al. (2013) recently suggested that biodegradation would be rapid and extensive when oil is present at concentrations expected with dispersant use. McFarlin et al. (2014) confirmed that Arctic microorganisms indigenous to Chukchi Sea water were capable of performing extensive biodegradation of chemically dispersed oil at an environmentally relevant temperature (-1°C) without nutrient additions. Indigenous microorganisms degraded both fresh and weathered oil at environmentally relevant concentrations, in both the presence and absence of dispersant, with oil losses ranging from 46% to 61% and up to 11% mineralization over 60 days without nutrient additions. When tested alone, 14% of 50 ppm of dispersant was mineralized within 60 days. Arctic species and their temperate region counterparts have been shown to have similar tolerance to dispersed oil, and the use of dispersant did not increase the toxicity of the oil (Gardiner et al., 2013).

There are at least two challenges to performing representative dispersed oil biodegradation studies in small-scale test systems (Lee et al., 2013). The first is how to conduct tests at the low dispersed oil concentrations that accurately represent field conditions. Many previous biodegradation studies were conducted in closed systems at unrealistically high concentrations (NRC, 2005). The second is how to maintain a stable dispersion in the laboratory. In order to reflect dispersion of oil in ocean conditions, biodegradation studies need formation of dispersed droplets of 70-100 microns and enough mixing energy to keep those droplets from resurfacing. This can be challenging in a closed system, especially during multiweek test periods, and demonstrates a weakness of many bench tests employed to study dispersant effectiveness and toxicity. They may accurately determine relative performance of different dispersants but are difficult to translate to real-world situations.

Formation of Oil-Mineral Aggregates In the marine environment, oil particles may not remain as discrete particles; they coalesce and rise back to the surface, or they can interact with suspended organic and/or inorganic particulate matter in the water column (Lee et al., 1985; Muschenheim and Lee, 2003; Owens and Lee, 2003) to form aggregate "flocs," which include oil-mineral aggregates (OMAs). The formation of OMAs stabilizes the oil-water interface, with the suspended particulate matter acting as a surfactant. This favors droplet formation and enhances oil dispersion into the water column. The process of stabilizing oil droplets with fine clay particles has led to natural and proactive oil spill remediation strategies for shorelines (Bragg and Yang, 1995; Lunel et al., 1996; Owens, 1999; Lee et al., 2003a). Breaking waves on the beach provide sufficient mixing energy to form OMAs from fine sediments and spilled oil (Lee et al., 2003a). The OMAs transport oil away from the shore, simultaneously providing a microcosm for rapid bacterial biodegradation (Lee et al., 1998b; Stoffyn-Egli and Lee, 2002). OMA formation enhances the natural dispersion of oil and reduces its environmental persistence (Bragg and Yang, 1995; Lee et al., 2003a; Owens and Lee, 2003; Owens et al., 2003). Since OMA formation results from the stabilization of oil droplets by mineral fines, a synergistic effect may be achieved with the addition of dispersants (Li et al., 2007).

Laboratory studies have shown that OMAs can quickly form at near-freezing temperatures in seawater if high-energy mixing is applied (Cloutier et al., 2005; Khelifa et al., 2005). These results were confirmed in a mesoscale basin containing brackish water with slush and broken ice, in which 20-30 minutes of mixing dispersed about 50% of the spilled oil (Blouin and Lee, 2007; Cloutier and Doyon, 2008). These trials were reproduced at full scale in January 2008 in the St. Lawrence River, when 200 liters of fuel oil were mixed with chalk fines by an icebreaker propeller (Lee et al., 2009a, 2011a). After dispersion into the water column, the oil did not resurface. A control test with no added particles produced significant resurfacing oil. Water samples taken back to the laboratory revealed that OMAs were formed and that more than half of the total petroleum hydrocarbons had degraded after 56 days of incubation at 0.5°C (Lee et al., 2012).

Promising New Response Concepts for Dispersion and Biodegradation

The *Deepwater Horizon* response demonstrated that large-scale subsea dispersant injection may be an effective approach for wellhead blowout spill mitigation. A major benefit of direct subsea dispersant injection is the ability to continuously respond independent of darkness, extreme temperatures, strong winds, rough seas, or the presence of ice. The dispersant volume could be substantially less than a surface application because of its ability to deliver dispersant very close to the oil release source, a key advantage with respect to enhancing oil encounter rate (how much oil a particular countermeasure can intercept or treat in a given time). This could be useful given the long and difficult logistics resupply chain in most Arctic areas, and, since less oil will come to the surface, subsea injection could lower volatile organic compounds encountered by responders at the sea surface. A comparison of response effectiveness showed that direct injection can disperse oil at rates significantly higher than those achievable by aerial dispersant application or other response methods (Federal Interagency Solutions Group, 2010). Another advantage of direct injection is higher efficiency, because the dispersant rapidly (within seconds) mixes with oil in a highly turbulent state at the discharge point. Because of the extremely rapid mixing under turbulent conditions, subsea injection is a potentially viable response even for wells in relatively shallow water, such as the Chukchi Sea. Based on research performed in a variety of mixing regimes (e.g., NRC, 2005; SL Ross Environmental Research Ltd. and MAR Inc., 2007, 2008; Reed et al., 2009), it is expected that a significant percentage of oil discharged from the Macondo well was converted to droplet sizes below 100 microns. Results from scaled-down laboratory experiments in vertical test tanks have demonstrated the potential generation of small oil droplets following dispersant additions to a subsurface discharge of oil at its point of release (Brandvik et al., 2013; Johansen et al., 2013). Still, more work needs to be done to understand the effectiveness, systems design, and short- and long-term impacts of subsea dispersant delivery. Rather than the ad hoc emergency injection system using remotely operated underwater vehicles and wands (as in the *Deepwater Horizon* response), future systems could involve pre-engineered injection lines directly into the well bore.

Mullin (2012) summarized a number of dispersant research studies undertaken by the current Arctic Response Technology JIP, while API and its industry companies have established a large-scale, multiple-year Subsea Dispersant Program, whose goal is to conduct controlled experiments on the effectiveness of subsea injection over various conditions, the effects of dispersed oil on deepwater marine environments, and numerical modeling needs for better prediction of oil fates.[12] This program recently released an initial report summarizing the status of dispersant regulatory approval and conditions on the application of dispersants in different Arctic nations (SEA Consulting Group, 2013).

Concerns over the resurfacing of oil dispersed under ice are also being addressed by the Arctic Response Technology JIP (Mullin, 2012). Scientists in Norway, the United States, and Canada are assessing whether turbulence levels in the water column of the Arctic Ocean are sufficient to keep oil suspended for a sufficient time for effective biodegradation to occur. Results are expected for public release and publication in peer-reviewed journals in late 2014.

For dispersant application in ice-covered waters, newer gel formulations that are more effective due to reduced solvent concentrations are currently being developed by industry. The gel formulation could increase the window of usability by being able to treat a wider range of crude oils and by increasing the amount of oil that may be treated by a given volume of dispersant. Future possibilities include spraying gel on oil that rises to the surface of the ice in the spring rather than igniting it,

[12] See www.api.org.

or adding gel dispersant to oil discharged from a surface blowout onto an ice cover (Nedwed et al., 2011).

Another recent effort involves the development of a jet aircraft–based dispersant delivery system that extends beyond the current use of Lockheed Hercules C-130As. This system was created in association with Oil Spill Response Limited at the recommendation of the International Association of Oil & Gas Producers' (OGP's) Global Industry Response Group. The Boeing 727 was selected because of its high transit speed, large payload, and extended range, which offers the possibility of effective response to spills in remote settings where other equipment may be less readily deployed (OGP, 2011; OGP and IPEICA, 2012). Aircraft are currently being configured for dispersant operation and will be put through an extensive number of tests to verify their performance.

Future approaches to enhance oil biodegradation include the application of nutrient-bearing treatment products (Kjeilen-Eilertsen et al., 2011) and the application of surfactants to surf-washing operations to increase the production of oil droplets and promote OMA formation. Another approach is bioaugmentation, where microbes with high biodegradation potentials are used to supplement existing microbial populations at contaminated sites (Bartha, 1986). This has often been proposed for use in situations where the indigenous microbial population cannot degrade petroleum, is stressed from recent exposure to an oil spill, has too small a population to maintain high biodegradation rates, or where seeding could reduce the lag period in growth prior to the onset of bioremediation (Forsyth et al., 1995). However, bioremediation field trials in open water environments have shown that bioaugmentation provides little or no benefit to treatment of spilled oil (K. Lee et al., 2005; Nichols and Venosa, 2008) and would not be expected to do so where natural seeps would have already enriched a larger number of adapted, oil-degrading microbes.

In Situ Burning

Controlled ISB of an oil slick as a response technology has been utilized for many years. While the first recorded use of ISB as a response countermeasure technique was in 1958 during a pipeline spill in the Mackenzie River, Northwest Territories (McLeod and McLeod, 1972), some important early work was carried out by the USCG in Alaska in the 1970s (McMinn, 1972). ISB is especially suited for use in the Arctic, where ice can often provide a natural barrier to maintain the necessary oil thicknesses for ignition, without the need for booms. A number of large-scale experiments successfully used ISB on oil that surfaced in spring melt pools after being spilled beneath the ice and trapped through a full winter. These experiments were carried out in the Canadian Beaufort Sea in 1975, 1980, and 1981, and in Svalbard in 2006 (NORCOR Engineering & Research Ltd., 1975; Dickins and Buist, 1981; Brandvik and Faksness, 2009). Overall removal rates ranged from 65% to greater than 90%, depending mainly on the size distribution of the melt pools. High efficiencies were documented for ISB of oil within fire-resistant booms in both Arctic ice-covered and southern open water environments (Potter et al., 2012).

ISB was used successfully on a trial basis during the *Exxon Valdez* response (Allen, 1990). In 1993, a U.S.-Canada experiment off Newfoundland burned crude oil in fire-resistant booms in the open ocean and monitored a large suite of environmental parameters including smoke composition, residue toxicity, and upper-water-column impacts (Fingas et al., 1995). The most recent Arctic experiences with the use of ISB offshore were through the Oil in Ice JIP (Sørstrøm et al., 2010), in which oil that was allowed to drift and weather in very close pack ice for over a week was successfully ignited and burned (Brandvik et al., 2010). Most recently, the large-scale ISB operation in response to the *Deepwater Horizon* blowout resulted in a unique set of full-scale operational data applicable to response planning for Arctic offshore areas in the summer. A total of 411 controlled burns removed an estimated 220,000 to 310,000 barrels of oil from the Gulf of Mexico (Allen et al., 2011).

Experience with burning fresh, weathered, and emulsified oils and petroleum products in a range of ice conditions has led to some basic "rules of thumb" (Buist et al., 2003a). The most important parameter that determines the likelihood of success and expected removal efficiency is the oil thickness, since below a minimum thickness of about 3 mm the burn will self-extinguish due to heat loss to the water. In order to achieve 60%-80% removal efficiency in most situations, the starting thickness of crude oil needs to be on the order of 3-5 mm. With relatively fresh oil that is wind herded against an ice edge or on melt pools in the spring, removal efficiencies in excess of 90% are achievable. Another key operational constraint is wind speed; a maximum for successful ignition is about 10 m/s or 20 kn, based on experience gained in field testing (Buist et al., 2003b). Concentrated pack ice can enable ISB by keeping oil slicks at the appropriate thickness (Buist and Dickins, 1987). In very open drift ice conditions, oil can spread rapidly and become too thin for ignition (Potter et al., 2012). Fire booms can collect and keep slicks thick in open water; however, even light ice conditions make the use of booms challenging (Bronson et al., 2002). In spite of these challenges, Potter and Buist (2010) reported highly effective (~90%) burning of oil within small ice pieces and brash collected within a fire-resistant boom during 2009 field experiments in the Norwegian Barents Sea. Ice concentrations in these tests were between 1/10 and 3/10, with small boats used to corral the needed quantities of ice. Conventional booming techniques are most effective in ice concentrations up to 1/10, with some effectiveness in concentrations of 2/10-3/10 (Potter et al., 2012). Despite these test results, there is continued concern among some non-governmental organizations that actual spill conditions could reduce the effectiveness of ISB to far below the theoretical maximum (e.g., WWF, 2010; Goodyear and Beach, 2012). In practice, experiences with very large burns at sea have demonstrated that efficiencies increase with scale, as the oil is pulled into the burn area by strong radial inflow winds at the surface (Buist et al., 1994). This was shown in observations during the *Deepwater Horizon* response, where free-floating, organized burning slicks could be seen following the oil corralled within fire booms.

In 2004, a multiyear joint industry and government (Minerals Management Service, now managed by BSEE) project began to study oil-herding chemicals to thicken slicks for ISB, as an alternative to booms in open drift ice conditions. Small-scale laboratory experiments were followed by midscale testing in large basins (e.g., Buist et al., 2007). The cold-water herder effectively thickened oil slicks in ice concentrations of up to 70%. Herded slicks in excess of 3 mm thickness were routinely achieved, and were ignited and burned at air temperatures as low as −17°C. The burn efficiencies that were measured for the herded slicks were only slightly lower than theoretical maximums achievable for slicks that are equivalently sized and physically contained on open water (Buist et al., 2011). The concept of using chemical herders to burn free-drifting oil slicks in pack ice was successfully field tested for the first time in the Norwegian Barents Sea in 2008 as part of a JIP on Oil Spill

Contingency for Arctic and Ice-Covered Waters (Buist et al., 2010). Burn removal effectiveness in that test was estimated to be on the order of 90%. The residue floated readily and was recovered manually from the water surface and ice edges. Buist et al. (2011) summarized past research into chemical herders and concluded that oil spill responders should consider utilizing them to enhance ISB in light to medium ice concentrations.

In the 1990s, research efforts assessed the potential environmental impacts of ISB, primarily from smoke plume and burn residues (Fingas et al., 1995). The smoke plume emitted by burning an oil slick on water is often the primary ISB concern to the public and regulators, as low concentrations of smoke particles at ground or sea level can persist for a few kilometers downwind. In practice, smoke particulates and gases are rather quickly diluted to concentrations below levels of concern (Fingas et al., 2001). Work by Canadian and U.S. teams advanced the understanding of smoke constituents and how to predict downwind environmental impacts (McGrattan et al., 1995). This work included a series of medium-scale burns at fire test facilities in Alabama, as well as a highly documented large-scale burn at sea off the Canadian East Coast in 1993—the Newfoundland Oil Burn Experiment (Fingas et al., 1995).

Burn residue—the unburned oil that remains on the surface of the water after a fire extinguishes naturally—was also studied in the 1990s. Daykin et al. (1994) and Blenkinsopp et al. (1997) studied burn residue's potential for aquatic toxicity, while an industry-funded research program examined the likelihood of burn residue sinking as it cooled (Buist et al., 1995; SL Ross Environmental Research Lt., 1998). Bioassays showed very little or no acute toxicity to oceanic organisms for either weathered oil or burn residue. These findings of little or no impact were validated with further studies by Gulec and Holdway (1999) and Gannon and Holdway (1999). Another area of historical research was to study the overall mass balance of PAHs consumed and created by ISB. During the Newfoundland Oil Burning Experiment, PAH concentrations were much lower in the plume and in particulate precipitation at ground level than they were in the initial oil composition, suggesting that PAHs are largely consumed by combustion (Fingas et al., 2001).

Numerous agencies, primarily in the United States, have established guidelines for the safe implementation of ISB as a countermeasure. The U.S. National Institute of Standards and Technology, NOAA, and Environment Canada have computer models used to predict safe distances for downwind smoke concentrations. In 1994, the Alaska Regional Response Team incorporated ISB guidelines for Alaska into its Unified Response Plan, becoming the first Arctic area to formally consider ISB as an oil spill countermeasure (ARRT, 2008). Their guidelines are considered the most fully developed to date. The American Society of Testing and Materials began developing standards associated with ISB in the late 1990s (ASTM, 2009), while the USCG produced an operations manual that details considerations and steps to be taken for open water ISB with fire booms (Buist et al., 2003a). New fire-resistant and fireproof boom designs were developed after a successful test burn during the *Exxon Valdez* spill (Allen, 1990, 1999). Several different types of fire booms were tested during the *Deepwater Horizon* oil spill, with some notable differences in their effectiveness for oil retention and durability in the face of fire intensity and sea state (Mabile, 2010).

All blowouts involve a mixture of oil and gas, mostly methane. While the casing of any wells drilled in the Chukchi or Beaufort OCS should extend below the shallow permafrost zones that could contain gas hydrates, ISB response strategies generally involve collecting and burning oil a safe distance away from the discharge site. This ensures that sea surface gas concentrations are well below the limit that could lead to spontaneous ignition and combustion.

Promising New Concepts for Improving ISB

The Helitorch was originally developed for the U.S. Forest Service to set deliberate fires and was adopted by oil spill responders in the 1980s as a means to ignite oil slicks at sea and on ice (Allen, 1987). From an operational perspective, Helitorch use may be appreciably constrained by weather conditions, darkness, aircraft icing conditions, and distance to the spill site. In the mid-1990s, new formulations for Helitorch fuel improved the ignition of emulsified and hard-to-light slicks. More recently, gelled delivery systems that are potentially capable of being operated at higher speeds from a fixed-wing aircraft were tested in ground trials (Preli et al., 2011). API is sponsoring a program with the U.S. Forest Service and other agencies to evaluate safer, more effective alternatives to the Helitorch. The Arctic Response Technology JIP has as one of its research priorities the development of new high-speed aerial ignition systems (Mullin, 2012).

Other new ISB projects include a BSEE-funded Naval Research Laboratory study on the use of atomizers to burn emulsified oil,[13] and validation and testing of an operational airborne application system for chemical herders planned under the Arctic Response Technology JIP (Mullin, 2012). In 2014, the JIP is also initiating a new project to evaluate the potential of chemical herders under different oil properties and weathering, as well as investigating windows of opportunity for their use.

Mechanical Containment and Recovery

Potter et al. (2012) define "containment and recovery" as actions taken to remove oil from the surface of water by containing the oil in a boom and/or recovering the oil with a skimming or direct suction device or sorbent material. After removal, the recovered mix of oil and water and contaminated materials needs to be stored offshore until it can be transferred to an approved disposal or recycling facility. Mechanical containment and recovery is often preferred over other oil spill countermeasures because they are viewed as directly removing oil from the marine environment.

However, there are problems using only mechanical containment and recovery, especially when dealing with large offshore spills in a remote area like the Arctic. Response to a large marine offshore spill is unlikely to rely only upon mechanical containment and recovery. Even though there were very good conditions for mechanical containment in the *Deepwater Horizon* spill, only an estimated 2% to 4% of the oil volume discharged was collected (Federal Interagency Solutions Group, 2010). This highlights a key drawback of mechanical containment and recovery systems when confronted by a large, rapidly spreading oil slick—there is an insufficient encounter rate to allow the skimmers to achieve a significant percentage of their theoretical recovery capacity (Allen, 1999). With any large spill in open water, oil rapidly spreads and forms a thin layer on the surface. Huge quantities of containment boom are needed to concentrate these thin slicks, requiring hundreds of vessels and skimmers to deploy and maintain. Skimmers can recover large volumes of water with the oil, result-

[13] See http://www.bsee.gov/Research-and-Training/Oil-Spill-Response-Research/Projects/Project1012/.

ing in serious storage and disposal requirements that would be challenging in the Arctic (S. Ross Environmental Research Ltd. et al., 2010). With smaller, more contained spills in harbors and other protected waters, mechanical containment and recovery can be very effective. However, very large oil spills require a response approach that does not solely depend on mechanical recovery.

The use of booms and skimmers during the Arctic summer months, with open water, follows the same procedures practiced for decades in more temperate regions. The Arctic has extended daylight in the summer, which provides one great benefit for operation. However, the lack of approved disposal sites on land in many Arctic coastal areas, lack of port facilities to accept deep-draft vessels, and limited airlift capability to remote communities complicate the large-scale use of mechanical containment and recovery to respond to spills. Conventional booms and skimmers become increasingly ineffective as ice concentrations increase. Limited effectiveness is possible in very open drift ice (1/10 to 3/10) and in isolated polynyas within closer pack ice. The presence of ice interferes with boom operation and reduces flow to the skimmer head, greatly reducing overall effectiveness (Bronson et al., 2002; Potter et al., 2012).

Past research programs such as Mechanical Oil Recovery in Ice Infested Waters (Mullin et al., 2003), supported by industry and the Minerals Management Service, evaluated the concept of mechanically processing ice pieces, flushing the oil off, and returning the cleaned ice to the ocean. While it was successful at a very small scale of a few meters, scaling up to the size of realistic ice floe sizes led to a massive piece of equipment able to work in only a limited set of Arctic conditions. Finland developed specialized ship-based systems that process oiled ice, but they are tailored to handle very small floe sizes typical of frequently used shipping channels in the winter (Lampela, 2007). Scaling such systems up to handle Arctic ice is probably impractical, and limited by low encounter rates.

Currently, improvement of mechanical systems for Arctic applications focuses on cold temperature protection (e.g., heating and water injection systems to pump viscous oil at cold temperatures, and heating and enclosing sensitive skimmer components) and independent propulsion (e.g., allowing skimmers tethered to a mother ship to be guided into oiled leads). The SINTEF Oil in Ice JIP tested a number of new Arctic skimmer prototypes in tanks and offshore field trials in 2008 and 2009 (Sørstrøm et al., 2010). Several of these prototypes are now available commercially.

In situations involving small, contained spills in pack ice or larger spills under fast ice, mechanical recovery can provide a viable option. In the Baltic Sea, isolated pockets of oil contained in close pack ice have been mechanically recovered with over-the-side brush- or bucket-type skimmers (Lampela, 2007). Oil spilled under fast ice and contained by its undersurface roughness can be recovered mechanically through cutting, trenching, and drilling (ACS, 2012). Depending on location, oil recovered in this manner can be transported to shore over smooth ice or on ice roads for disposal and/or reinjection. These proven response strategies are most applicable to locations such as Prudhoe Bay, where stable fast ice extends tens of kilometers offshore.

Promising New Concepts in Mechanical Recovery

Future improvements in conventional mechanical recovery systems are likely to be incremental rather than transformative (Sørstrøm et al., 2010). The fundamental constraint of limited encounter rate is made harder by the presence of ice and is not easily overcome with existing or proposed systems, although further modifications and improvements in cold-temperature operability are likely to continue. Although not generally considered in the category of mechanical recovery, well-capping systems can be viewed as a crucial component of mechanical recovery. They may be the most effective mechanical means of quickly stopping oil discharge from a subsea blowout, and they theoretically eliminate the need for any subsequent response.

Nonetheless, efforts have been undertaken to encourage the improvement of conventional mechanical recovery systems. One such effort was the Wendy Schmidt Oil Cleanup X Challenge,[14] a competition designed to inspire innovative solutions that speed the pace of cleaning up marine oil spills from ships and other sources. The one-year competition ended in fall 2011, with the prize awarded to two teams that demonstrated oil recovery rates of over 2,500 gallons/minute, with efficiency rates of more than 70%. In 2014, BSEE is also planning to fund projects that potentially increase mechanical recovery of oil in drift ice.

Issues with Operational Limits and Recovery Rates of Countermeasures

The knowledge gained from laboratory, tank, and field experiments under Arctic conditions can be used to determine operating limits for different countermeasures (Potter et al., 2012). However, operating limits by themselves are not good indicators of response effectiveness. For example, while it may be possible to ignite an oil slick in winds up to 10 m/s, the response will not necessarily achieve a high removal rate. Instead, removal rate will depend on a number of factors, such as oil thickness and degree of emulsification. Similarly, it is possible to deploy skimmers into isolated oil pools trapped between floes, but the recovery rate may be relatively low due to a need to continually relocate the skimmer and potential interference of ice with the skimmer intake (Bronson et al., 2002; Potter et al., 2012). Dispersants can theoretically be applied over a wide range of conditions, but unless a sufficient dose consistently reaches the thicker areas of an oil slick, overall effectiveness may be low.

Encounter rate may be the most critical factor for oil removal. This will depend on the speed of advance (e.g., vessels that tow boom, aircraft speed), the swath or sweep width (e.g., boom opening, aircraft or vessel spray width), the burn removal rate, and skimmer recovery rate. Allen et al. (1999) provides an in-depth discussion of the encounter rate as it applies to different response options. Calculating the expected recovery or removal rate for a particular response effort is a complex process, with no simple means of estimating spill response effectiveness. Instead, past experience with particular spills can help to assign ranges of recovery effectiveness for specific countermeasures. Net environmental benefit analysis (NEBA) can help responders, regulators, and stakeholders decide which oil spill response options could be recommended or advised against in a given situation. The principles of applying this type of analysis are discussed in Chapter 5.

As seen in the sections above, there are continuing efforts by industry and government to improve oil spill response both in the Arctic and elsewhere, but concerns remain regarding the ability to effectively mitigate a large spill in the Arctic or other remote locations. Members of the nongovernmental organization community maintain that even when used in conjunction, mechanical

[14] See http://www.xprize.org/content/wendy-schmidt-oil-cleanup-x-challenge.

recovery, ISB, and biodegradation and dispersants may still only remove a small fraction of spilled oil (Goodyear and Beach, 2012; presentations by Stan Senner, Ocean Conservancy; Peter Van Tuyn, the Pew Charitable Trusts; and Chris Krenz, Oceana, June 2013). Furthermore, these groups call for rigorous testing under a range of Arctic conditions before using these techniques during a major spill.

Detection and Monitoring

To mount an effective response, it is critical to know where spilled oil is at any given time. Finding and mapping oil in ice, even under favorable weather and light conditions, is far from straightforward. The lack of significant waves in the presence of ice complicates the use of marine or satellite radar systems, both of which depend on differences in surface roughness as a means of detection. With adverse weather or darkness, obtaining consistently reliable detection and mapping of contaminated areas becomes challenging and requires a mix of remote sensors operating in different parts of the electromagnetic spectrum. Assessments of remote sensing system capabilities for oil spills in ice have been forced to draw upon practical detection experiences of spills in open water environments (Coolbaugh, 2008; Dickins and Andersen, 2009; Fingas and Brown, 2011).

A wide range of sensor types have been tested for use in spill detection in ice. These include analytical, bench, and basin tests and field trials using a wide range of sensor types—acoustics, radar, ultraviolet fluorescence, infrared (IR), gamma ray, microwave radiometer, resonance scattering theory, gas sniffers, and ground penetrating radar (GPR) (e.g., Dickins, 2000; Goodman, 2008). Beginning in 2004, projects sponsored by the Minerals Management Service and industry (including the SINTEF Oil in Ice JIP) evaluated and tested a variety of sensors currently used to detect oil on open water to evaluate their potential for detecting oil in ice. In addition to the sensors above, these projects looked at side-looking airborne radar, synthetic aperture radar (SAR) satellites, forward-looking infrared (FLIR), trained dogs, and sonar (Bradford et al., 2010; Dickins et al., 2010). The current Arctic Response Technology JIP (2012-2015) is examining a range of these and other airborne, surface, and subsurface technologies in order to assign priorities for future development and testing (Puestow et al., 2013; Wilkinson et al., 2013).

Table 3.4 compares the capabilities of different sensors for remote sensing of oil spills in ice according to the platform and the oil/ice configuration over a range of ice environments (Dickins and Andersen, 2009). SINTEF JIP field experiments in 2008 and 2009 evaluated some of these technologies in close pack ice. Expected capabilities of different systems are based on information gathered during those experiments and from results of previous trials, not necessarily in the Arctic.

Dickins and Andersen (2009) concluded that current airborne systems are useful for detecting and mapping large spills in open ice but have much less potential as ice increases. Many of the non-radar sensors on airborne systems do not work well under Arctic conditions of darkness, cloudiness, fogginess, and rain for much of the year. A quantum leap in all-weather capability was realized in the late 1990s with the advent of commercially available, high-resolution SAR satellite systems, which are unaffected by darkness or cloud cover and can now resolve targets of a few meters (e.g., Radarsat, ERS-1, TerraSAR-X, COSMO-Skymed). First-generation SAR satellites mapped several large marine oil spills, including the *Prestige*, *Nakodka*, and *Sea Empress* (Hodgins et al., 1996; Lunel et al., 1997). The ability of SAR satellites to detect and map oils slicks in the ocean with moderate wind conditions is likely to be practical for well-defined oil spills that spread in very open to open pack ice, where capillary waves can develop on the surface (Babiker et al., 2010).

The *Deepwater Horizon* oil spill provided an opportunity to utilize many of the latest detection technologies. Leifer et al. (2012) summarized how passive and active satellite and airborne marine remote sensing were applied to the spill. Slick thickness and oil-to-water emulsion ratios are key spill response parameters for containment and cleanup. These parameters were derived for thick (greater than 0.1 mm) slicks detected by the Airborne Visible/Infrared Imaging Spectrometer satellite. The Moderate Resolution Imaging Spectroradiometer (MODIS) satellite data allowed for detection of the total slick and was used to produce maps of oil thickness. Airborne and SAR provided synoptic data under all-sky conditions; however, SAR typically is not able to discriminate between thick oil slicks and thin sheens (0.1 mm or less). The Jet Propulsion Laboratory's Uninhabited Aerial Vehicle SAR's[15] higher spatial resolution and signal-to-noise ratio led to better pattern discrimination.

At present, there is a lack of hard data to confirm theoretical assumptions about the performance of most remote sensing systems in a particular oil-in-ice scenario (GPR, discussed below, is one exception). SAR satellite imagery may be of use for detecting oil slicks in ice but will be dependent on factors such as the size of the spill, ice floe size and concentration, and wind speed. The main value of radar imagery in an Arctic spill incident lies in its ability to document changing ice conditions near a spill, which provides a valuable tactical tool for response (Potter et al., 2012). False positives or negatives are a concern with SAR imagery, because other conditions can create smooth surface areas that resemble oil (e.g., grease ice), while atmospheric phenomena can hide potential oil slicks (e.g., rain cells).

An important national issue is that the United States does not presently have its own commercial SAR satellite mission, so international partnerships are necessary. Access to images from international satellites typically only lasts about three days, which could be an issue for oil spill response (presentation by Nettie LaBelle-Hamer, University of Alaska Fairbanks, March 19, 2013). Classified SAR satellites from the National Geospatial-Intelligence Agency can be used for monitoring natural disasters and oil spills, but the information is only available to those people with appropriate clearances. However, derivative maps can be made using the data. Having emergency responders able to access necessary SAR data is of great importance for an integrated response.

In addition to rapidly developing remote sensing technologies, there will always be a need for well-trained observers flying in helicopters and fixed-wing aircraft to map oiled areas and to transmit critical information to response crews. Spotter aircraft were essential to the success of individual ISB operations during *Deepwater Horizon* (Allen et al., 2011).

Comprehensive tracking and long-term monitoring of oil released in ice require assimilating field data, plotting real-time observations, and integrating this information with forecasting tools such as weather models, ice drift algorithms, and oil spreading and weathering models. Commercially available ice-strengthened Global Positioning System (GPS) beacons and buoys have, for many years, been tracking ice movements during an entire winter season throughout the polar basin (Vaudrey and Dickins, 1996). Tracking oil spills accurately in a moving ice cover involves deploying large numbers

[15] See http://uavsar.jpl.nasa.gov/.

Table 3.4 Overview of Remote Sensing Systems for Detection of Oil in Ice

Platform	Ice Surface		AUV	Shipborne	
Sensor	Dogs	GPR	Sonar	Marine Radar	FLIR
OIL ON ICE					
Exposed on cold ice surface	Likely	NA	NA	Not likely	Likely*
Exposed on spring melt pools	Likely	NA	NA	Possible	Likely*
Buried under snow	Likely	Likely	NA	NA	Not likely*
OIL UNDER ICE					
Smooth fast ice	Possible	Likely	Likely	NA	NA
Deformed pack ice	Possible	Possible	Likely	NA	NA
OIL IN ICE					
Discrete encapsulated layer	Possible	Likely	Possible	NA	NA
Diffuse vertical saturation	Possible	Possible	Not likely	NA	NA
OIL BETWEEN ICE FLOES					
1/10 to 3/10 concentration	NA	NA	Not likely	Likely	Likely*
4/10 to 6/10 concentration	Not likely	NA	Not likely	Possible	Likely*
7/10 to 9/10 concentration	Possible	NA	Not likely	Not likely	Likely*

NOTE: An asterisk (*) denotes sensors blocked by dark/cloud/fog/precipitation.
SOURCE: Dickins and Andersen (2009).

Airborne					Satellite
GPR	Visible	UV	FLIR	SLAR	SAR
Likely	Likely*	Not likely*	Likely*	Not likely	Not likely
Not likely	Likely*	Possible*	Likely*	Possible	Not likely
Likely	Not likely*	Not likely*	Not likely*	Not likely	Not likely
Likely	NA	NA	NA	Not likely	Not likely
Possible	NA	NA	NA	Not likely	Not likely
Likely	NA	NA	NA	Not likely	Not likely
Possible	NA	NA	NA	Not likely	Not likely
Not likely	Likely*	Likely*	Likely*	Likely	Likely
Not likely	Likely*	Possible*	Likely*	Possible	Possible
Not likely	Likely*	Not likely	Likely*	Not likely	Not likely

of beacons at regular intervals on the ice as oil moves away from the spill source, which can then be used to direct air and marine responders toward the spill. Closely spaced GPS beacons can follow the evolving pattern of spill fragmentation and divergence as the pack expands and contracts (Hirvi et al., 1987; Weingartner et al., 1998). There are also subsurface Lagrangian floats that can operate in ice-covered waters and be acoustically tracked while under sea ice, with capabilities including subsurface profiles down to 2,000 m water depth. These will be deployed under the Office of Naval Research Departmental Research Initiative Marginal Ice Zone Program.[16] In the absence of sea ice or in reduced sea ice cover, the floats surface and send water-column profiles and subsurface position data via satellite; two-way communications allow the float's drift depth and/or profiling frequency to be changed when the float surfaces.

Promising New Concepts in Detection and Monitoring

A number of systems, including GPR, are capable of both airborne and surface operation. Tank and field experiments from 2004 to 2006 showed that surface-based, commercially available GPR can detect and map oil sheens as thin as 1-3 cm underneath 1 m or more of solid ice or trapped as layers within ice (Dickins et al., 2008). In 2008, the same radar suspended beneath a helicopter traveling at speeds up to 20 kn and altitude up to 20 m successfully detected a thin layer of crude oil buried under hard-packed snow (Bradford et al., 2010). A prototype frequency-modulated continuous-wave radar designed to detect oil trapped under solid ice from a low-flying helicopter was developed in 2011-2012, with plans to test it as part of a new Arctic Response Technology JIP research program beginning in 2014.

More recently, consideration is being given to utilizing nuclear magnetic resonance as a potential means to detect oil trapped under or in ice in the future (Nedwed et al., 2008). Although further testing is needed to evaluate the practicality and effectiveness of an operational system, nuclear magnetic resonance has been successfully used to locate groundwater on multiple occasions.

IR systems (alone or in conjunction with high-speed marine radar and low-light-level video) can also be used from the surface, low-flying helicopters, aircraft (tracking high-resolution FLIR), or vessels. In 2009 SINTEF JIP tests, a basic uncooled hand-held IR sensor was able to discriminate between oil, open water, snow, and oil-free ice floes during daytime (Dickins et al., 2010). These experiments confirmed prior offshore releases in pack ice, in which IR differentiated warmer oil from cold water and ice (Singsaas et al., 1994).

Depending on ice conditions (e.g., floe size, thickness, stability), a variety of remote sensing systems that operate directly from the ice surface or from a nearby vessel could be deployed. As part of the SINTEF JIP, trained dogs on the ice tracked and located small oil spills buried under snow from a distance of 5 km and also determined the approximate dimensions of a larger oil spill (Brandvik and Buvik, 2009).

X-band marine radar has been used to detect slicks at sea in large-scale trials, and may be able to detect oil slicks in open ice (Dickins and Andersen, 2009). Integrated systems that combine high-

[16] See http://www.apl.washington.edu/project/project.php?id=miz.

resolution FLIR and low-light cameras are now routinely deployed on response vessels in Norway as the SECurus system.[17]

Unmanned aerial vehicles (UAVs) and autonomous underwater vehicles (AUVs) already have the capability of carrying useful sensor packages over long distances (albeit at slow speed) for Arctic oil spill surveillance (Wadhams et al., 2006). Both single- and multi-beam sonar sensors successfully detected and mapped oiled boundaries and thicknesses under ice in a recent basin test at the U.S. Army Cold Regions Research and Engineering Laboratory (Wilkinson et al., 2014). A September 2013 exercise aboard the USCG *Healey* field tested UAVs, AUVs, and Arctic skimmers for response capabilities (USCG, 2013a). Further testing of different UAV sensors for oil spilled in ice is planned for 2014 and is being sponsored by the European Union and Arctic Response Technology JIP.

BSEE has recently partnered with the Army Research Development and Engineering Command to develop new sensing capabilities that could have applications during low-light periods,[18] which could be especially useful for work in the Arctic.

Arctic nations such as Canada, Iceland, Finland, Denmark, Norway, and Sweden operate dedicated pollution surveillance aircraft. Canada dedicates one of its aircraft (a DHC-7) to Arctic missions. From 1983 to 2013, the USCG operated a fleet of Falcon Jets for maritime surveillance; most of these obsolete HU-25s have been decommissioned, with complete phase-out by 2014.[19] They are being replaced by the HC-144A Ocean Sentries, which entered service in 2009 (there are currently 15). The search radars in USCG fixed-wing aircraft have a SAR setting that can be used for oil spill detection. There was some limited evaluation during *Deepwater Horizon*, but no rigorous testing has been performed and no testing with ice has yet occurred. The fixed-wing aircraft and helicopters have an electro-optical/IR system that may be useful in some Arctic conditions (Kurt Hansen, USCG, personal communication, February 3, 2014). A key aspect of the future effectiveness of remote sensing systems is the ability to integrate different datasets into a useful real-time or near-real-time product that responders can use with minimal interpretation. While there has been considerable progress on multispectral data fusion applied to pollution surveillance aircraft, an equivalent capability for Arctic spill surveillance has yet to be developed (Baschek, 2007). Lessons learned from *Deepwater Horizon* (Leifer et al., 2012) could be applied to optimizing future Arctic systems for detecting oil in ice.

Trajectory Modeling

Several reviews of oil spill modeling technology are available (Huang, 1983; Spaulding, 1988; ASCE, 1996; Reed et al., 1999; Yapa et al., 2006; Drozdowski et al., 2011). Both earlier and more recent reviews recognize the challenges associated with oil spills in Arctic conditions—the presence of sea ice, cold and darkness in the winter, sparse observational networks for meteorological and oceanographic conditions, and a limited ability to respond to and monitor the physical and chemical evolution of a spill.

[17] See www.aptomar.com.

[18] See http://www.bsee.gov/Research-and-Training/Oil-Spill-Response-Research/Projects/Project1013/.

[19] See http://www.uscg.mil/hq/cg7/cg711/hu25.asp.

Oil in ice behaves in complex ways. Challenges associated with modeling the physics of ice growth, movement, and deformation on scales of meters or tens of meters are magnified when the details of oil behavior are incorporated. Background literature exists on oil-ice interaction studies over the past 40 years. DF Dickins Associates Ltd. and Fleet Technology Ltd. (1992) and SL Ross Environmental Research Ltd. et al. (2010) provide overviews of key studies. More recent work has largely focused on oil in and under ice (Yapa and Belaskas, 1993; El-Tahan and Venkatesh, 1994; Yapa and Weerasuriya, 1997), but mostly relies on small-scale, short-term laboratory studies. Ice lead dynamics that govern the spreading of oil in the field tend to not be included in these solutions, a significant shortcoming.

Understanding the effect of Arctic conditions on oil spill behavior and fate has increased significantly over the past decade (e.g., Gjøsteen et al., 2003; Faksness, 2007; Dickins et al., 2008; Brandvik and Faksness, 2009; Buist et al., 2009; Brandvik et al., 2010). Much of this knowledge has been garnered during mesoscale and field experiments (e.g., Singsaas et al., 1994; Buist et al., 2007; Sørstrøm et al., 2010), which show that in ice leads, evaporation, dispersion, and emulsification are slowed. The primary factors in determining observed weathering rates appear to be temperature, wave damping, and the presence of ice.

Despite these advances, there continues to be a need for additional Arctic data and for incorporation of these data into comprehensive oil spill models (Holland-Bartels and Pierce, 2011). A key problem in achieving such integration lies in a limited ability to model ice behavior at appropriate (meters) spatial scales. Forecasting (Reed and Aamo, 1994) and hindcasting (Johansen and Skognes, 1995) demonstrate the issues that can be encountered when oil-ice interaction models are used in the field. The limited capabilities of modeling ice behavior at scales of 1-10 m also significantly restrict the extent to which advances in oil spill modeling can be used. Ice coverage is dynamic and can change rapidly, with significant implications for oil weathering and transport.

Gjøsteen (2004) produced a model for spreading of oil in irregularly shaped simulated ice fields. Russian work in this field exists but is often not published in English, restricting its use and citation (e.g., Ovsienko et al., 1999). Both Gjøsteen and Ovsienko have developed spreading models that account for spreading of oil among ice floes. Incorporation into numerical models of these advances, as well as increased understanding of oil weathering processes in the presence of sea ice (e.g., Brandvik et al., 2004, 2005, 2010; Dickins et al., 2008; Faksness, 2007; Brandvik and Faksness, 2009; Buist et al., 2009), has been hindered by the interdisciplinary nature of the problem. Integration of knowledge of oil behavior and fates, ice cover, and hydrodynamic models is necessary for further significant advances in oil-ice interaction modeling.

Achieving higher spatial resolution using existing classic sea ice models (e.g., Hibler, 1979) is not sufficient, as robust oil spill models will need more detailed representations of sea ice (e.g., ice floe sizes, ice porosity, ice drift, ridging and growth rates, under-ice roughness). Oil spill models can then compute and retain fractions of oil on, in, and under ice floes, and allow for dynamic partitioning in both space and time. Advancement in this direction is needed for both sea ice and oil spill models, although neither is likely to be able to reliably perform at this level of detail in the near future.

The Arctic Oil Spill Response Technology JIP has recently initiated a research effort to improve oil spill trajectory modeling capability within the Arctic, with plans to develop new sea ice models related to ice dynamics. The models will be evaluated at regional scales using a high-resolution, coupled ice-ocean model, and a new type of model will be developed to simulate sea ice dynamics in the marginal ice zone. All of the products developed during the JIP will be integrated into established oil spill trajectory models (Mullin, 2012).

BOEM currently uses a model called the Oil Spill Risk Analysis Model,[20] which simulates the possible paths of oil spills through modeled ocean currents, winds, and sea ice. In March 2011, a workshop was held with the goal of improving BOEM oil spill models (SAIC, 2011). A main recommendation was the need for assimilating new oceanic, atmospheric, and sea ice data collected in the Arctic environment. Another was the requirement for validation and sensitivity testing of the model through integration with observational studies—for example, under programs such as the Forum for Arctic Ocean Modeling and Observational Synthesis.[21]

NOAA's Office of Response and Restoration's Emergency Response Division also employs an oil spill trajectory model, which is effectively a statistical model that is run many times to determine an uncertainty boundary where oil may be found. The General NOAA Operational Modeling Environment[22] predicts possible trajectories of oil in water for given environmental forcing. An Arctic module contains information on tides, currents, bathymetry, and coastlines in the Arctic region, but does not include sea ice.

In 2013, BSEE and NOAA entered into an interagency agreement to adapt NOAA's current models to better account for Arctic conditions. Their efforts are intended to complement the work being done by others, including the Arctic Oil Spill Response Technology JIP.

Promising New Concepts in Trajectory Modeling

Promising advances in sea ice modeling in the past decade include detailed models of brine-channel formation and drainage by Petrich et al. (2006, 2013). This approach allows for incorporation of oil into brine channels as well as bulk oil freezing into ice (Faksness and Brandvik, 2008; Faksness et al., 2011). Hopkins (1996) and Hopkins and Daly (2003) have developed a discrete element approach to modeling sea ice that allows for variably sized ice floes. These two advances together permit a parameterization of oil-ice interactions at a conceptual resolution that is significantly closer to reality than was previously possible. Wilkinson et al. (2007) have demonstrated the possibility of modeling the flow of oil under sea ice based on the topography of the under-ice surface. The authors relied on an AUV to map the under-ice contours, an approach that is probably not realistic over a long time frame and a dynamic ice field. Future ice models will need to produce an estimate of under-ice roughness if the spreading process is to be adequately represented. Other advances in oil spill modeling are occurring, although they are driven mostly by the *Deepwater Horizon* rather than by issues associated with the Arctic (e.g., the Gulf of Mexico Research Initiative[23]). There is a strong focus on underwater near-field plume modeling, including the effectiveness of dispersant injection at the wellhead.

[20] See http://www.boem.gov/Environmental-Stewardship/Environmental-Assessment/Oil-Spill-Modeling/Oil-Spill-Risk-Analysis-Model-%28OSRAM%29.aspx.

[21] See http://www.whoi.edu/projects/famos/.

[22] See http://response.restoration.noaa.gov/oil-and-chemical-spills/oil-spills/response-tools/gnome.html.

[23] See http://gulfresearchinitiative.org/.

Uncertainty as Related to Numerical Tools for Decision Support

Numerical simulation models of coupled atmospheric, sea ice, and ocean physics are key components of today's weather, ice, and ocean current forecasting operations. The success of these tools in providing accurate forecasts is defined by the degree to which model results reflect reality. This, in turn, depends on an understanding of the underlying processes, building this understanding into the models, and on available computational power. Simulation tools are also used for planning and training, support during a spill response, and evaluation of potential and actual environmental effects after an accident. These tools bring additional physical, chemical, and biological processes into the equation, along with their associated uncertainties.

Reliable forecast tools depend heavily on the assimilation of real-world observations because numerical models are by definition simplified representations of reality, relevant processes and their linkages are incompletely understood, and there are computational limitations on the ability to represent these processes over the full range of physical and temporal scales. The ability to supply observational data will remain a key element in reducing this uncertainty. For weather forecasts, such observations typically include wind speed, air temperature, humidity, and other environmental variables that are measured regularly at numerous locations on the Earth's surface and in the overlying atmosphere. The density of similar measurements in the oceans and at the air-ocean interface is much less, but has increased significantly over the past two to three decades, particularly at lower latitudes. Such measurements are scarcer in the Arctic, where the complexity of the physical relationships is increased by seasonal growth, drift, and decay of sea ice. Adding spilled oil increases this complexity.

Simulation models are powerful integrators of complex interacting processes and events, and can be useful tools for decision support in both planning and executing emergency response actions. Statistical and ensemble analyses are two standard approaches to quantify the uncertainty of model results.

Statistical modeling is typically used in planning for a possible future event, when details of the event are not known beforehand. In oil spill contingency planning, for example, the timing, durations, and magnitudes of possible releases from a given production facility are unknown but can be estimated within certain bounds. One can then run a large number of simulations sampling from ranges of start times, release rates, and durations to produce probability maps for a variety of end points—for example, oil arrival time, the amount of oil on the sea surface, or the amount of oil along shorelines. Results can be used to support planning that protects priority resources. The probability of an impact on a given resource, weighted by the probable magnitude of the impacts, supplies measures of environmental risk that can in turn supply criteria for electing alternative response options.

Ensemble modeling simulates a given event using multiple models, then interprets the variability of the resulting outputs as a measure of intrinsic model-related uncertainty. This approach was used by both circulation and trajectory models during the *Deepwater Horizon* oil spill (e.g., Chang et al., 2011; Marianoa et al., 2011). It is also commonly applied in climate change modeling (e.g., Djalalova et al., 2010) and is most useful when there are multiple available models of similar complexity and focus.

Uncertainty in oil spill trajectory forecasting for open water conditions has been a recognized challenge for decades (e.g., Galt, 1994) and will probably be greater in the presence of sea ice (Eicken, 2013). Trajectory analyses for decision support will need to account for this uncertainty in ways that are useful to response practitioners. Overlaying the trajectory forecast and the associated envelope of uncertainty on natural resource maps provides a graphical basis for decision making and reflects the present state of the art (e.g., NOAA Gulf Spill Restoration, 2012).

CHAPTER CONCLUSIONS AND RECOMMENDATIONS

Conclusion: Arctic oil spill research and development needs for improved decision support include:

- Determining and verifying biodegradation rates for hydrocarbons in offshore environments, in order to establish potential and capacities for natural attenuation or recovery and to determine which strategies can accelerate oil biodegradation;
- Improving technologies for dispersant application and induction of turbulence for oil spills in ice;
- Evaluating the toxicity of dispersants and chemically dispersed oil on key Arctic marine species, with appropriate experimental design and incorporation of real-world conditions and concentrations;
- Identifying and understanding ecosystem responses associated with changes in microbial biomass and species alterations;
- Further understanding of biogeochemical cycles, including particulate matter–petroleum chemical interactions;
- Improving ignition methods for in situ burning, focused specifically on the Arctic;
- Mapping the usefulness of chemical herders at different spatial scales, oil types, and weathered states, and in conjunction with other response options such as in situ burning;
- Understanding the limitations of mechanical recovery in both open water and ice, which will improve decision making regarding possible implementation of other response strategies;
- Investing in under-ice detection and response strategies, including remote sensing technology that will reliably detect oil in different ice conditions;
- Integrating remote sensing and observational techniques for detecting and tracking ice and oil, including UAVs, AUVs, SAR, and drifting buoys;
- Additional research into the physics of oil incorporation into developing ice;
- Establishing robust operational U.S. Arctic meteorological-oceanographic-ice and oil spill trajectory forecasting models for contingency planning and response support;
- Testing and evaluating risk-benefit decision processes, including NEBA, for use in the Arctic; and
- Summarizing relevant ongoing and planned research worldwide to achieve synergy and avoid unnecessary duplication.

Conclusion: Though much is known about the oil behavior and response technologies in ice-covered environments, there are areas where additional research is needed to make informed decisions about the most effective response strategies for different Arctic situations. In addition, there is a need to validate current and emerging oil spill response technologies on operational scales under realistic environmental conditions.

Conclusion: A systematic program of carefully planned and controlled field experiments that release oil in the U.S. Arctic is needed to advance understanding of oil behavior and response options.

Recommendation: A comprehensive, collaborative, long-term Arctic oil spill research and development program needs to be established. The program should focus on understanding oil spill behavior in the Arctic marine environment, including the relationship between oil and sea ice formation, transport, and fate. It should include assessment of oil spill response technologies and logistics, improvements to forecasting models and associated data needs, and controlled field releases under realistic conditions for research purposes. Industry, academic, government, non-governmental, grassroots, and international efforts should be integrated into the program, with a focus on peer review and transparency. An interagency permit approval process that will enable researchers to plan and execute deliberate releases in U.S. waters is also needed.

Recommendation: Priorities for oil spill research should leverage existing joint agreements and be addressed through a comprehensive, coordinated effort that links industry, government, academia, international and local experts, and non-governmental organizations. The Interagency Coordinating Committee on Oil Pollution Research, which is tasked to coordinate oil spill research and development among agencies and other partners, should lead the effort.

Conclusion: The oil spill response toolbox requires flexibility to evaluate and apply multiple response options, whether on their own or concurrently. No single technique will apply in all situations.

Recommendation: Dispersant pre-authorization in Alaska should be based on sound science, including research on fates and effects of chemically dispersed oil in the Arctic environment, experiments using oils that are representative of those in the Arctic, toxicity tests of chemically dispersed oil at realistic concentrations and exposures, and the use of representative microbial and lower-trophic benthic and pelagic Arctic species at appropriate temperatures and salinities.

Conclusion: Well-defined and tested decision processes for oil spill countermeasure deployment are critical to expedite review and approval. Decision processes need to include rapid research on countermeasures and be exercised regularly.

4 Operations, Logistics, and Coordination in an Arctic Oil Spill

Several types of commercial activities are increasing in the Arctic, leading to the prospect of rapid growth in shipping along several routes. For example, use of the Northern Sea Route, and to a lesser extent the Northwest Passage, as a transportation route (Figure 4.1) is now more possible than ever before (IPCC, 2014). While some commercial shippers do not believe it will be economically viable for shipping in the near future (presentation by Gene Brooks, Maersk, February 2013), increased seasonal use by tankers and tug barges seems likely (Arctic Council, 2009). Taken along with other forms of vessel traffic, such as the tanker traffic from the Northern Sea Route, bulk carriers and tug-barge traffic transporting minerals and other bulk commodities, the inevitable increase in fishing fleets as fish stocks migrate northward, and even cruise ships that offer a glimpse of the Arctic for tourists, the Arctic has become a much busier place, with all of the associated risks that increased traffic involves. For the United States, the implications for traffic management, and by association, environmental protection, are very real. Keeping oil out of the water will not be purely a function of sound drilling practices, but of sound vessel traffic management, which raises a host of concerns for protection of Arctic ecosystems and for preparedness to respond in the event of a marine accident. It is also a concern for all Arctic nations, as an oil spill that occurs in one part of the Arctic may cross geographic boundaries and impact other nations' waters, food supplies, livelihoods, and cultural resources.

Spills from these anticipated activities are likely to be relatively small and involve lighter oils (e.g., diesel, heating oil). Less likely but more consequential spills would be associated with offshore oil exploration and production, as well as from large bulk carriers operating from Kotzebue. Vessels operating from U.S. ports are subject to the contingency planning requirements of the Oil Pollution Act of 1990 (OPA 90). To cope with changing traffic patterns in the Arctic, significant investments in infrastructure and capabilities, such as navigational aids, charting, communications, real-time traffic monitoring, ice forecasting, ship repair, and salvage capabilities, are needed.

As first introduced in the risk-based framework of Chapter 3, seven oil spill response scenarios are presented throughout this chapter and the next. Each scenario is color-coded to match its location on the risk-based framework shown in Figure 3.1. There are a number of issues that are common to all the Arctic scenarios in the report. For instance, oil spill response strategies will vary with

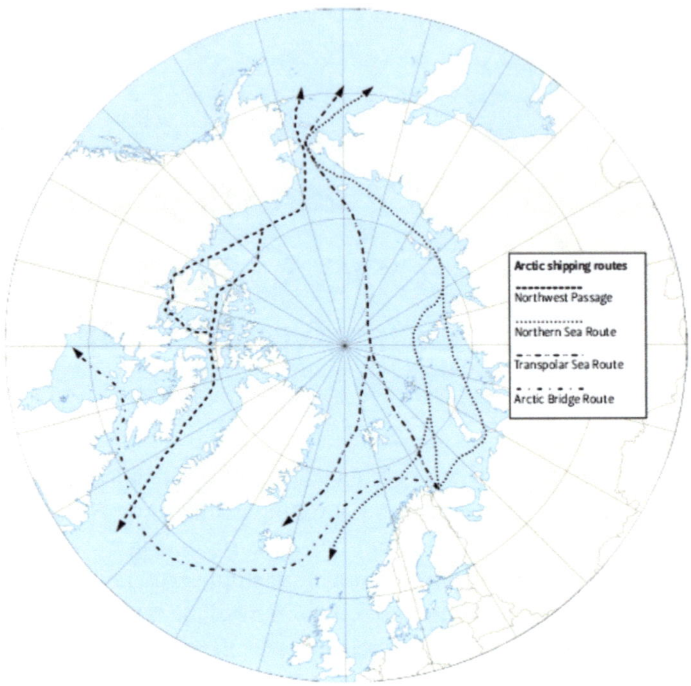

Figure 4.1 Arctic sea routes. SOURCE: Mikkola and Käplyä (2013); data from the Arctic Institute.

weather, ice cover, oil type, location, and available resources. A critical issue is the general inability to immediately respond and a lack of supporting infrastructure and capabilities (e.g., equipment, training, logistics), which has repercussions on response effectiveness. In some instances, especially in international events, the determination of who is responsible for the spill or who has authority for a response may need to be clarified. The issue of a responsible party, discussed later in this chapter, may impact how response, recovery, and salvage will be carried out and how eventual environmental restitution will be funded. Further issues of importance are the effects on the social, economic, and subsistence patterns on Arctic communities and impacts to the environment and ecosystem.

RESPONSE ORGANIZATIONS

Viable spill response options will vary depending on the nature of the spill, its location, and prevailing environmental conditions. Other variables to be considered include the proximity of the oil spill to sensitive marine ecosystems; the presence and density of marine life; and, in many coastal Arctic areas, culturally significant sites (ADEC Spill Prevention and Response, 2002b). Each spill requires an assessment of the risks and efficacy of potential response options in order to remove the pollutant as completely as possible without causing additional environmental harm. Not all response

Scenario 1
Passenger Cruise Ship Accident

A passenger cruise ship accident in the Arctic could occur from a ship grounding due to poorly charted bathymetry, operator error, or ice, among other possibilities. This could occur anywhere along the coastline of the U.S. Beaufort or Chukchi Seas.

Main Considerations
The ship may be foreign flagged and/or may not have intended to enter the U.S. Exclusive Economic Zone, but the accident and potential spill will have impacts in U.S. waters and may require U.S. resources. Search and rescue operations would take precedence over oil spill response throughout the event, and logistics such as housing, food, medical support, and sanitary resources in remote Arctic villages for rescued passengers and responders could be difficult. Personnel rescue and/or recovery could consume scarce resources that are available and planned for oil spill response.

Standard Response
Responsible party (tasked to assume rescue and oil spill response leadership) is identified, Unified Command is activated, and (as mentioned above) personnel rescue is prioritized over oil spill response.

Response Needs
High-quality nautical charts, ice data and charts, adequate training and exercises for combined or complex incidents (personnel rescue/evacuation/oil spill response), and supply chain logistics for rescued passengers and oil spill responders are needed.

tools are appropriate in every scenario, but each should be considered within the framework of a Net Environmental Benefit Analysis, which will be considered more completely in Chapter 5.

Managing these variables requires a multiskilled response organization that can rapidly assess the situation, develop response priorities, identify and deploy response resources, conduct removal operations, and monitor the results. These activities need to be informed by scientific knowledge, benefits and drawbacks of each option need to be properly weighted, and activities need to be undertaken with an understanding of the complex legal framework that imposes responsibilities on different parties in an oil spill response.

Marine oil spills can occur from a number of sources and under a wide range of conditions, including originating from land sources or from vessels. A vessel spill that threatens U.S. Arctic waters and coastlines can originate in waters beyond U.S. jurisdiction. In such a case, the response organization will take on an international dimension, involving a broader and more complicated coordination challenge.

OPA 90 places the primary responsibility for mounting an effective response on the spiller (known as the responsible party [RP]). However, federal, state, local, and tribal entities each have legal authorities to oversee, approve, and, if needed, supplement the measures undertaken by the spiller. A wide range of federal and state laws, as well as international treaties and agreements, weaves a complex web of duties, responsibilities, and authorities for all parties involved in a spill response. Accounting for these various laws and agreements is an integral aspect of a response organization's operations. Consequently, U.S. spill response efforts are organized along a standard construct that is followed by federal and state officials and by operators of vessels and regulated marine facilities.

The National Response System

The National Response System is the federal government's mechanism for responding to oil spills and releases of hazardous materials. It operates through a network of federal agencies, described in the regulation under which EPA and USCG conduct spill response, the National Oil and Hazardous Substance Pollution Contingency Plan (40 C.F.R. Part 300).

Key Components

Key components of the National Response System and its relevance for the Arctic include the following:

Federal On-Scene Coordinator (FOSC)—a federal official designated to coordinate the federal government's response to a spill. The FOSC also coordinates with the affected state(s) and directs the response efforts of the responsible party. Although the law places the principal response burden on the RP, the FOSC may supplement or direct the RP's actions if needed to ensure an effective response. In cases where the RP is not known, the FOSC will directly manage the response using funds from the Oil Spill Liability Trust Fund (OSLTF). In the coastal zone, the predesignated FOSC is typically the U.S. Coast Guard (USCG) Captain of the Port, while the Environmental Protection Agency (EPA) is predesignated for a spill in the inland zone. For the Arctic coastal zone, jurisdiction lies with the Western Alaska USCG Captain of the Port. There are also State On-Scene Coordinators (SOSCs) for a response; in Alaska, the Alaska Department of Environmental Conservation (ADEC) serves as the predesignated SOSC.

Area Contingency Plan—FOSCs are also responsible for working with the Area Committee, a group comprising federal and state representatives, to develop an Area Contingency Plan that identifies sensitive areas and resources to be protected. The plan also identifies likely pollution scenarios, available response options, and the logistical requirements to put the plan into effect. Participation by local officials and non-governmental organizations is encouraged, as is participation by commercial spill response organizations. The State of Alaska is a member of the Area Committee and is represented by ADEC. Area committees will also normally consider special response techniques that may be appropriate for their geographic region.

In Alaska, the Area Contingency Plan requirement is satisfied through the *Alaska Federal and*

State Preparedness Plan for Response to Oil and Hazardous Materials Discharges and Releases[1] (known as the Unified Plan), which was developed jointly between the State of Alaska, USCG, and EPA. The Unified Plan contains 10 geographically specific subarea plans, which allow for focused planning and interaction with local officials and tribal leaders (ADEC Spill Prevention and Response, 2010a). In the Arctic, these are the North Slope and Northwest Arctic SubArea Plans.

Regional Response Team (RRT)—a team comprising federal, state, and local officials with broader geographic responsibility than an individual FOSC, charged with providing advice and assistance. RRTs are typically co-chaired by the EPA Regional Office and the USCG District Office. RRTs are directly involved in decisions related to the use of special response techniques, such as dispersants and in situ burning (ISB), including pre-authorization for their use under certain defined conditions.

The Alaska RRT is co-chaired by EPA Region 10 and the USCG 17th District. Federal resource trustees from the National Oceanic and Atmospheric Administration (NOAA) and the U.S. Department of the Interior are members, as are a number of agency representatives in the region. While ISB is pre-authorized under defined conditions, the Alaska RRT has not granted pre-authorization for the use of dispersants.

National Response Team (NRT)—a team comprising 16 federal agencies with expertise in environmental protection and response. It is a planning and coordination body that produces national-level policy. Although not envisioned as a response body, the NRT can provide real-time assistance and advice to decision makers in the event of a Spill of National Significance (discussed below).

National Response Center—a 24-hour national call center for reporting oil spills and hazardous materials releases. Spillers are required to notify the National Response Center even if they have already notified local officials. The National Response Center ensures that all relevant federal agencies and the appropriate FOSC are promptly advised of the report so that an appropriate response can be initiated.

Special Teams

National Strike Force—three specialized teams that are maintained by the USCG but are available to any FOSC and are available to deploy on short notice. They possess equipment and training that exceed typical FOSC resources. Teams are located in California (Pacific Strike Team), New Jersey (Atlantic Strike Team), and Alabama (Gulf Strike Team), but they regularly support responses outside of their primary areas as needed. When deployed, they work directly for the FOSC and can fit into a standard response organization.

Environmental Response Team—an EPA team of specially trained scientists and engineers who have expertise in sampling, analysis, hazard assessment, and technical support.

[1] See http://dec.alaska.gov/spar/perp/plans/uc.htm.

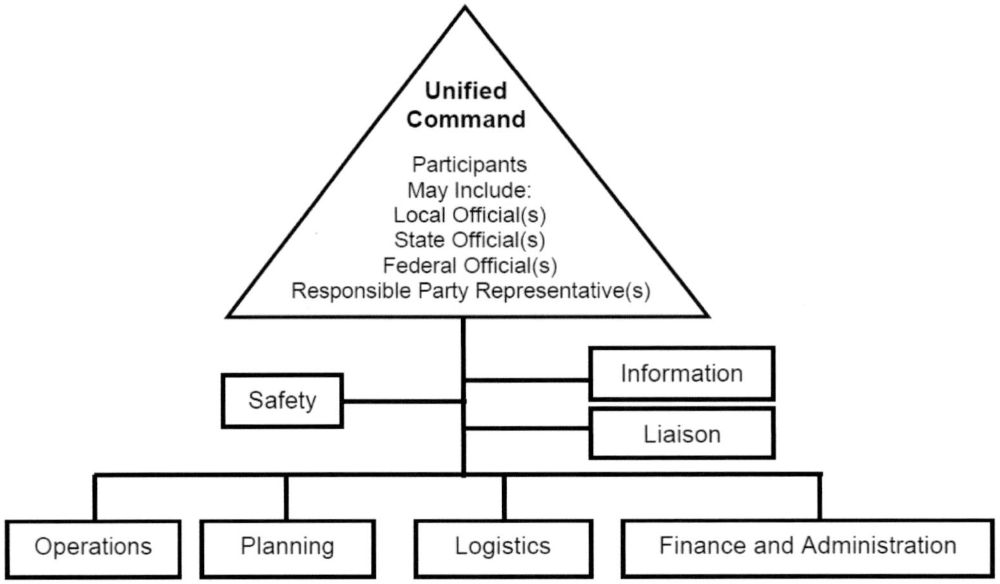

Figure 4.2 Relationship between Incident Command System and Unified Command.
SOURCE: National Response Team Technical Assistance Document: http://www.nrt.org/Production/NRT/NRTWeb.nsf/AllAttachmentsByTitle/SA-52ICSUCTA/$File/ICSUCTA.pdf?OpenElement.

Scientific Support Coordinator—an individual assigned to assist the FOSC in gathering and analyzing environmental and safety information in order to enable the FOSC to make timely operational decisions. For spills in the coastal zone, scientific support coordinators are typically provided by NOAA.

National Incident Management System/Incident Command System

Spill response involves multiple organizations coming together to address a common problem. Some of these organizations may never have worked together before, although responders in the same geographic area are likely to have coordinated through periodic exercises designed to test the Area Plan. The Federal Emergency Management Agency (FEMA) National Response Framework envisions a National Incident Management System/Incident Command System (NIMS/ICS)-based response (Figure 4.2), which originated in the fire service to coordinate multiple firefighting services for fighting western wildfires.[2] The ICS concept is now used by the federal government to organize for all emergencies (DHS, 2008). NIMS/ICS includes standard organizational structures, planning cycles, and terminologies, and enables all participants (the RP and federal, state, local, and tribal governments) to organize in a consistent, commonly understood manner. It provides a unified command structure, with a command element typically comprising the FOSC, SOSC, and the RP's

[2] See http://training.fema.gov/EMIWeb/is/ICSResource/assets/reviewMaterials.pdf.

On-Scene Coordinator. In Alaska, there is also a provision for a Local On-Scene Coordinator to represent local communities. No one surrenders their authorities or responsibilities under this construct; instead, they are positioned to carry out their responsibilities in a coordinated way. Exercises that employ ICS principles are conducted on a regular basis so that federal and state responders are familiar with each other. There is active involvement by the communities and tribal leaders, who provide invaluable local knowledge and expertise.

The *Alaska Incident Management System Guide for Oil and Hazardous Substance Response* provides standardized oil spill response management guidelines to responders in Alaska (ADEC, 2002). The *Alaska Incident Management System Guide* meshes with the National Response Framework and NIMS/ICS but is specific to the state's interests.

LEGAL FRAMEWORK FOR POLLUTION PREPAREDNESS AND RESPONSE

The response organization must fulfill its function in compliance with a host of federal and state laws, international treaties and agreements, and contingency plans. These are detailed below.

U.S. Federal Laws and Regulations

The Oil Pollution Act of 1990

The Oil Pollution Act of 1990 (OPA 90; P.L. 101-380), possibly the most well-known federal pollution statute, was passed in the wake of the *Exxon Valdez* spill in Prince William Sound, Alaska. It is the principal statute underpinning preparedness requirements, response organization, and liabilities of the responsible party and has had a profound effect on the way the United States responds to marine spills.

OPA 90 amended and strengthened provisions of the Clean Water Act (P.L. 92-500). While OPA 90 reinforced the concept that the RP has the primary duty to organize and carry out an effective response, it enhanced the authority of the FOSC to direct the response actions. The law also imposed strict planning requirements on vessel operators, including securing the equipment needed to effectively respond to spills and identifying a qualified individual that has the authority to commit company resources to respond to a spill or potential spill. OPA 90 also mandated that commercial operators have the financial ability to fund a spill response originating from their facility or vessel. How a vessel or marine facility intends to respond to spills of various magnitudes is contained in a formal Response Plan, which must receive federal approval before the vessel or facility is allowed to operate. Not all RPs are equal in terms of their liability limits or degree of preparation, as noted below:

- **Tank vessels** are subject to the most stringent OPA 90 requirements in terms of preplanning, vessel response plans, equipment, and exercise programs. Tankers must also demonstrate that they have liability coverage to fund their spill response costs and any associated environmental damage. Liability coverage is documented in a "Certificate of Financial

Responsibility" issued by the National Pollution Funds Center, an entity established by OPA 90, and administered by the USCG. The amount of required coverage varies by the gross tonnage, the type of vessel, and whether it is fitted with a double hull. Double-hulled tankers above 3,000 gross tons must provide coverage of $2,000 per gross ton or $17.088 million, whichever is greater (see 33 U.S.C. § 2704). Single-hull tankers would have an even greater liability—$3,200 per gross ton or $23.5 million, whichever is greater. Of note, single-hulled tankers are rapidly being phased out. All must have double hulls by 2014 due to OPA 90 requirements and comparable international standards in the International Convention for the Prevention of Pollution from Ships (MARPOL). There are no cases on record where response costs have exceeded liability limits for double-hulled tankers, but it is possible that if a large double-hulled tanker had a serious incident, response costs could quickly outstrip the liability limits. In the event of a spill, once the liability limit is reached, it may be difficult to compel the RP to continue funding the response (USCG, 2012).

- **Nontank vessels** (e.g., freight ships, cruise ships) are also subject to OPA 90 planning and response requirements and must also possess evidence of financial liability. However, specific planning details which would form the basis for the Vessel Response Plan have not yet been finalized in regulation. This is the subject of an ongoing regulatory project by the USCG, and represents additional variability in the preparedness of these vessels in the interim. Nontank vessel financial responsibility for vessels greater than 300 gross tons is $1,000 per gross ton or $854,400, whichever is greater. Given the potential for large cargo vessels to carry in excess of 400,000 gallons of fuel oil, their liability limit could conceivably be quickly reached in the aftermath of a serious casualty in the Arctic,[3] an issue of serious concern as shipping potentially increases.
- **Tank barges** such as those used in coastal fuel delivery are subject to similar planning and preparedness requirements as tankers. Their liability limits mirror those for tankers, based upon their tonnage and whether they are single or double hulled. Double-hulled barges of 3,000 gross tons or greater will have the same limits as for double-hulled tankers. However, many tank barges in regular service to the Alaskan Arctic are less than 3,000 gross tons. Assuming a double hull, their liability limit is $2,000 per ton or $4,272,000, whichever is greater.
- **Offshore facilities** are not limited in liability for spills, with the exception of deepwater ports. The liability is for all removal costs plus $75 million for damages.[4] Oil Spill Response Plans are reviewed and approved by the Bureau of Safety and Environmental Enforcement (BSEE) and are coordinated with subarea plans.
- **Onshore facilities and deepwater ports** have a liability limit of $350 million.[5] OPA 90 identified the NIMS/ICS as the standard response framework that plan holders must incorporate into their response organizations in order to facilitate the assimilation of multiple

[3] According to the USCG *Annual Report to Congress on Oil Spill Liability in 2012*, "the available data continue to suggest that the existing liability limits for certain vessel types notably tank barges and cargo vessels with substantial fuel oil, may not sufficiently account for historic costs incurred as a result of an oil discharge from these vessel types."

[4] 33 U.S.C. § 2704.

[5] 33 U.S.C. § 2704.

Scenario 2
Large Tanker Spill

Late in the season, a large tanker transiting through Russian waters from the Northern Sea Route experiences structural damage and loses power as it passes through the Bering Strait. After communicating with both Russian and American authorities, the tanker grounds on Big Diomede Island. The tanker is carrying nearly 30 million gallons of crude oil. Although much of the oil being carried is spilled immediately, some remains and continues to slowly seep out. The tanker is not owned by a U.S. individual or corporation, nor is it registered in the United States. During its transit, the tanker never entered U.S. water. However, the oil is likely to have significant impacts in U.S. waters, due to seasonal winds and currents.

Main Considerations
The primary issue is how to address a spill by a foreign flagged vessel on a voyage that never enters U.S. waters, yet could have major impacts on the United States due to the spilled oil. Bad weather, low visibility, ice impacts on the response and rescue efforts, shoreline contamination in remote areas, and oil in ice are likely to extend cleanup over the winter months.

Standard Response
In the absence of a clear or immediately identifiable responsible party, the U.S. Coast Guard will assume a leadership and oversight role. The Unified Command and U.S. Coast Guard may also exercise existing bilateral and multilateral international agreements.

Response Needs
Enhanced vessel traffic monitoring, protocols to identify the responsible party, adequate predeployed spill response equipment, resources for shoreline contamination cleanup, and adequate training and exercises for local response personnel (working with Unified Command personnel) in adverse weather during the winter are needed.

federal, state, local, and private entities into the response effort. Federal officials are also required to plan for their role in coordinating the federal response to a spill, in conjunction with state and local officials.

Executive Order 12777, issued October 1991, clarified some responsibilities assigned in OPA 90. This included that the National Contingency Plan (NCP) provide for an NRT and RRTs composed of representatives of appropriate federal departments and agencies for national or regional coordination and planning. Five agencies (Departments of Commerce, Interior, Defense, Agriculture, and Energy) were identified for inclusion in the NCP as trustees to look after federal environmental interests, while EPA and USCG were assigned to designate Areas and Area Committee Members. Responsibility for oversight of spill planning and equipment inspections was also assigned to the Coast Guard for ships and marine facilities, to EPA for shore-based facilities, to the Department of the Interior for offshore installations, and to the Department of Transportation for pipelines.

Scenario 3
Bulk Ore Carrier Driven Onshore in Bad Weather

In a scenario similar to the *Selendang Ayu*, a bulk ore carrier is in the Arctic during the summer, when shipments are being made from the Red Dog Mine. Storm conditions cause the ship to lose propulsion and ground it onshore. The bulk ore carrier is carrying heavy fuel oil in excess of 100,000 gallons.

Main Considerations
A major issue in this scenario is that while it is relatively easy to identify the bulk ore shipper as the responsible party, there is less regulatory oversight than there is with oil/fuel shippers. Other considerations include the potential for an oil spill due to loss of fuel, and that spilled or damaged cargo or its cleanup may have environmental or human health risks.

Standard Response
Responsible party assumes leadership role for cleanup, an oil spill removal organization (OSRO) is identified, and salvage capabilities and response are identified and activated.

Response Needs
Adequate Arctic oil spill response infrastructure, parity in regulatory oversight for bulk and petroleum shipping, identification of OSROs for oil spills from nonpetroleum shippers, resources for nearshore spill response, training and exercises for combined or complex events (salvage/cargo cleanup/oil spill response), and training and exercises for local responders working with the Unified Command are needed.

National Contingency Plan

While the Clean Water Act as amended by OPA 90 sets expectations and legal responsibilities, the overall details for how the government responds are contained in the NCP (40 C.F.R. Part 300), a regulation administered by the EPA. It was first developed in 1968 after the *Torrey Canyon* oil spill in the United Kingdom, in order to ensure that the United States was prepared to respond to spills in national waters. The NCP was updated and expanded after passage of the Clean Water Act in 1972 and OPA 90.

The NCP provides the regulatory framework for the NRT, RRT, and the role of the FOSC. It also establishes the unified command structure for handling responses and maps out the general pattern of response to be followed by the FOSC, including determining the threat, classifying the size and type of the release, notifying the RRT and National Response Center, and supervising removal actions.

The NCP defines Spills of National Significance (SONS) as having magnitudes that outstrip regional capabilities. These spills can have far-reaching public health effects, widespread environmental impacts, and economic damage. A SONS will involve senior levels of government and require response on a national scale and may also require the use of national capabilities not commonly envisioned in more routine responses—for example, national security assets such as satellite capability or military capabilities including naval, aviation, and command and control assets. The 2010 *Deepwater Horizon* oil spill was the first SONS to occur after OPA 90 was passed. SONS exercises are held every three years in different regions of the country and include an industry volunteer to serve as the RP.

Other National Legislation

In addition to OPA 90, there are several key laws that influence the nature of any response, the considerations for restoration, and the liability of the spiller. One of these is the National Environmental Policy Act (1970), which imposes requirements on federal agencies to conduct an environmental assessment and, if necessary, an environmental impact statement, for major federal actions affecting environmental quality. Typically, pollution response operations have been categorically excluded if there were no extraordinary circumstances. Another is the Act to Prevent Pollution from Ships (1980), which incorporates the MARPOL Convention into U.S. law (33 U.S.C. §§ 1901-1912). Other laws include the Endangered Species Act (1973), the Marine Mammal Protection Act (1972), the Migratory Bird Treaty Act (1918), and the National Historic Preservation Act (1966, as amended).

State Laws

ADEC is authorized to establish regulations that control, prevent, and mitigate all forms of pollution. Legislation enacted in 1980 defined Alaska's policies regarding oil spills (1980 Session Laws of Alaska, Ch. 116: "An Act relating to the prevention and control of oil pollution; and providing for an effective date"). The law's purpose is to ensure the safety and protection of human health and welfare of Alaskans from damage resulting from oil spills and to allow for cleanup and restoration of damaged areas following a spill. In 1989, after the *Exxon Valdez* oil spill, the Alaska Legislature enacted additional legislation to further strengthen the state's capability to deal with oil spills. The law was again revised in 1990 with specific elements, including oil discharge contingency plans, an incident command system that details specific responsibilities, and other actions to lessen oil spills and their consequences.

International Treaties and Agreements

MARPOL

MARPOL was originally adopted by the International Maritime Organization in 1973 and, following the MARPOL Protocol in 1978, came into force in 1983. The International Maritime Organization is a specialized body of the United Nations that has jurisdiction over maritime safety and marine environmental protection matters. MARPOL specifies design, equipment, and operational measures to prevent pollution, as well as the steps required in the event of an incident. Annex I (Oil) is specifically focused on oil pollution. In the 1980s, MARPOL was modified to require double hulls on tankers, closely matching the requirements contained in OPA 90 and in 2006, it was again modified to require enhanced protection for fuel tanks on nontankers because of their potential to cause major oil spills. The United States, Canada, and the Russian Federation are party to Annex I of this Convention, along with approximately 150 other nations, representing 99% of global shipping (IMO, in press). In 1980, the Act to Prevent Pollution from Ships was passed by Congress, giving MARPOL the force of law in waters under U.S. jurisdiction. All ships subject to the convention and bound for a U.S. port are required to comply.

International Convention on Oil Pollution, Preparedness, Response and Cooperation

The 1990 International Maritime Organization Convention on Oil Pollution Preparedness, Response, and Cooperation (OPRC) provides a global framework for international cooperation in the event of major spills, including planning and mutual assistance, regional preparedness, and research and development. OPRC provides an international construct for more specific agreements between nations, as exists between the United States and Canada, and between the United States and Russia. OPRC also requires that international ships have a Shipboard Oil Pollution Emergency Plan, detailed under the MARPOL convention, which contains the steps ships must take to notify the appropriate authorities and sources of assistance in the event of an emergency. However, OPA 90 requires even more detailed response plans than what is required by OPRC/MARPOL.

Other Treaties and Legislation

The International Convention Relating to Intervention on the High Seas in Cases of Oil Pollution Casualties (1969) affirms the right of coastal nations to take appropriate measures to prevent, mitigate, or eliminate damage to its coastline as a result of a marine casualty occurring beyond its territorial sea. This convention is reflected in U.S. law through the Intervention on the High Seas Act of 1974 (33 U.S.C. § 1471). The International Convention on Salvage (1989) imposes obligations on vessel owners and salvors to exercise due care to protect the marine environment. It addresses economic incentives for the salvor to prevent environmental damage and provides for increased cooperation between salvors, public authorities, and other interested parties.

The Canadian Arctic Shipping Pollution Prevention Regulations specify when and where certain categories of vessel can enter specific Shipping Safety Control Zones,[6] which cover Canadian Arctic waters from the U.S. border to Greenland. Zones were originally selected on the basis of ice severity, including dates of freeze-up or clearing and the potential to encounter multiyear ice. An amendment to these regulations now allows ship masters to enter and exit zones outside of the traditional zone/date system, but clearance to continue transiting is based on the characteristics of the ice regime and that vessel and ice conditions are transmitted to the Canadian Coast Guard. This system of controls applies to ships carrying more than 453 m^3 of oil, which encompasses most commercial vessels. The system's purpose is to prevent substandard vessels from entering Arctic

[6] See http://www.tc.gc.ca/eng/marinesafety/tp-tp12259-appendicies-2872.htm.

waters and posing a risk of spills through hull damage or ice penetration of fuel tanks. There is no equivalent mechanism at the U.S. state or federal level designed to achieve this level of risk reduction.

Diplomatic Agreements

In May 2013, the member states of the Arctic Council (Canada, Denmark [also representing Greenland and the Faroe Islands], Finland, Iceland, Norway, the Russian Federation, Sweden, and the United States) signed an Agreement on Cooperation on Marine Oil Pollution Preparedness and Response in the Arctic. Its objective is to strengthen cooperation, coordination, and mutual assistance regarding preparedness and response capacity for oil spills in the Arctic. The parties agree to share information on response techniques and to assist each other with technical capability and expertise to the extent of their ability should a spill occur. Additionally, the agreement dictates that each party will maintain a national system for responding to oil spills, which at a minimum would include a national contingency plan. The agreement also specifies how or to what extent each party will designate authorities and contact points, provide notification to other states, undertake monitoring activities, and provide assistance or request assistance from other parties.

Bilateral Agreements

The United States has long engaged its regional neighbors in Arctic spill preparedness. Bilateral agreements with Canada and Russia predate the more recent and much broader Arctic Council agreement on spill response among all arctic nations. The 1974 Canada-United States Joint Maritime Pollution Contingency Plan contains a framework for cooperation between the United States and Canada in the event of a spill affecting or potentially affecting both countries. The relationship with Canada has been particularly well developed through a formal Canada–United States Joint Marine Pollution Contingency Plan (CANUS) and associated exercises. It began as an initiative for the Great Lakes but soon expanded to cover boundary areas in the Atlantic, in the Pacific Northwest, and in Alaska. Each geographic region has a specific annex that is exercised on a regular schedule to ensure familiarity with the plans. The plans envision a joint response effort, where technical expertise and capability are joined in a coordinated effort. Broad governmental cooperation on each side of the border has enabled both countries to refine their procedures and legal requirements for cross-border movement of technical experts and equipment in the event of an emergency. CANUS exercises have generally been focused on the Dixon Entrance, rather than in the Arctic, but the agreement was recently tested in an Arctic setting through the binational CANUSNORTH environmental response exercise held in August 2012 in Tuktoyaktuk, Northwest Territories.[7] However, if oil exploration proceeds in the Canadian Beaufort Sea, a well blowout would likely transport oil westward into U.S. waters. In July 2013, the Canadian Coast Guard and USCG conducted their first-ever offshore oil spill drill near the Bering Strait, during which they deployed containment boom.[8] There have been

[7] See http://www.dfompo.gc.ca/media/npress-communique/2012/ca08-eng.htm.

[8] See http://www.alaskajournal.com/Alaska-Journal-of-Commerce/July-Issue-4-2013/US-Canada-conduct-Bering-Strait-spill-drill/.

other CANUSNORTH exercises, but most have been tabletop exercises to test policies and working relationships rather than equipment deployment.

The 1989 Agreement Concerning Cooperation in Combating Pollution in the Bering and Chukchi Seas in Emergency Situations focuses on the potential for both the U.S. and the Russian Federation to be affected by a marine spill in the Bering Sea, Bering Strait, and/or Chukchi Sea. It provides for the establishment of a joint response team and a joint response center in the event of an emergency. The agreement is administered by the USCG and the Russian Marine Pollution Control and Salvage Administration. Although there is communication between the respective responder agencies of the United States and Russia and oil spill communication exercises have been held, exercises to test the details of how such a large-scale response might occur have not occurred (RADM Thomas Ostebo and 17th Coast Guard District personnel, personal communication, February 4, 2013; March 20, 2013; and March 18, 2014). Given the increase in vessel traffic through the Bering Strait from ships using the Northern Sea Route (Marine Exchange of Alaska, 2009-2012), it is more likely that a joint response will be needed. The bilateral agreement provides a solid foundation for a joint response effort, although the myriad details of response cooperation during any given spill response could lead to significant delays unless worked out in advance.

If an oil spill occurred in Russian waters and was positioned to affect U.S. waters, it is essential for the United States to know where Russia would establish its command post, which of its agencies would be involved, what capabilities they would have available, and vice versa. There would also be a need for a structured information exchange to coordinate planning between the countries, including a mechanism to have liaison officers from each country at each other's command post if desired (which would require a spill response expert in both countries who are fluent in both Russian and English). Because there have not been joint exercises, it is unclear whether U.S. spill response resources would be allowed to enter Russian waters to assist and, if so, what clearances or agreements would be needed, and what controls would be imposed on traffic transiting in the area. However, there is a history of cooperation between the Kamchatka Border Guard and USCG District 17, including fisheries enforcement and port calls. In addition the USCG and the Russian State Maritime Pollution Control, Salvage, and Rescue Administration have agreements in place to facilitate search and rescue and pollution response. On a practical level, D17 staff communicate directly with their Russian counterparts about search and rescue alerts and other emergencies (McConnell, 2013).

There are also concerns regarding how the RP construct would work in this case. It is unknown whether the RP identified by Russia could be held responsible for environmental or socioeconomic damages in the United States, and, if the RP is not willing to fund response operations on the U.S. side of the strait, whether the OSLTF would be sufficient to cover open ocean response and/or potential sociological costs for loss of livelihood for subsistence hunters.

Discussion of Issues

For spills occurring within U.S. jurisdiction, OPA 90 provides the necessary legal framework for funding response operations and providing compensation for damages. OPA 90 places the burden of responding and for financing response activities on the RP. The RP's financial liability also extends to funding economic claims for those affected by the spill, and for natural resource damages. Potential RPs who regularly operate vessels within the Arctic region are likely to be well prepared, due to their OPA 90 responsibilities. However, situations could arise where OPA 90 may not in itself provide sufficient cover for those affected by a spill. Spills occurring outside of U.S. jurisdiction, but which affect resources in the U.S. Exclusive Economic Zone or territorial seas, are problematic. In such cases, and absent any significant assets held within the United States, the RP may be outside the reach of U.S. law. Those who are simply transiting through the region without conducting a U.S. port call (and thereby triggering OPA 90 preparedness requirements) are also likely to place a significant burden on government capabilities. There would be a need to mobilize open ocean response capabilities, utilizing commercial contracts and quite likely government resources. The U.S. Department of Defense (DOD) resources may be particularly useful, including communications capability, cargo aircraft, surface vessels, and specialized equipment maintained by the Navy Supervisor of Salvage. The Coast Guard, as the predesignated FOSC for the coastal zone, has been in contact with the Alaska Command regarding contingency operations and has worked out mechanisms by which appropriate DOD or Alaska National Guard assets could be requested if deemed essential to a response effort.

Without an RP, the costs incurred by the responders are borne by the OSLTF, a revolving fund maintained by the federal government and administered by the USCG. The OSLTF is maintained through recouping costs and penalties from responsible parties and by imposing a tax on oil (currently 8 cents per barrel) as authorized by Congress. The OSLTF is divided into two funds—the principal fund is used to pay claims and to fund appropriations authorized by Congress for federal agencies to administer OPA 90, while the emergency fund provides $50 million annually for use in removal actions and to initiate natural resource damage assessments. Legislative authority exists to increase this amount to $100 million if needed. Further increases require congressional approval, with a statutory limit of $1 billion per incident, provided there are sufficient funds remaining in the OSLTF. Given the distances and austere environment of the Arctic, a sustained response would be very expensive and could quickly draw down available OSLTF funding. In addition to direct response costs, there are often serious economic effects as well as potential health effects on local communities, which may also be chargeable to the OSLTF.

Unlike oil spill response, many domestic emergencies (such as hurricanes) make use of DOD assets after a Stafford Act declaration by the President. These assets are funded through FEMA after the issuance of Mission Assignments. A "whole of government" response, similar to a Stafford Act response, could be needed to manage community needs beyond direct economic loss (e.g., medical monitoring, seafood safety). This would not only incur additional costs, but also require expertise in multiple disciplines. During the *Deepwater Horizon* oil spill, a Stafford Act–like response organization needed to be overlaid onto the OPA 90 response effort to deal with the sociological aspects of the spill. Because there was no Stafford Act declaration, the effort was carried out under OPA 90 and was funded by BP, the RP in this incident.[9] It is conceivable that a similar hybrid response

[9] A responsible party whose expenditures exceed their limits of liability could seek reimbursement from the Principal Fund: in that sense, it remains to be seen what the full effects of the *Deepwater Horizon* will be on the OSLTF. BP established a $20 billion fund to pay claims—far in excess of the statutory $75 million.

organization would be needed in a future spill, particularly a SONS; however, there does not appear to be any mechanism to fund a hybrid response in cases where the RP is less financially capable.

The *Deepwater Horizon* response highlighted disconnects between the federal perspective of a Unified Command response under the NCP and state perspectives about state-led emergency response under the Stafford Act. These disconnects were operationally resolved during the *Deepwater Horizon* event by adoption of the NCP framework, but the potential for future confusion remains because of overlapping jurisdictions and ambiguity in responsibility (USCG, 2011). For Stafford Act–like responses, the OSLTF is probably not a sustainable source of funds. Moreover, the organizational structure under the National Contingency Plan is geared toward spill response. The broader social impacts resulting from a large spill may require a different structure for long-term management. Even though the population numbers in the Arctic are small, the native communities are to a far greater degree dependent upon the quality of their environment. A significant spill can have long-lasting effects on the social fabric of their communities.

ORGANIZATIONS AND INFRASTRUCTURE FOR OIL SPILL RESPONSE

Oil Spill Removal Organization Classifications

Section 4202 of OPA 90 amended the Clean Water Act to require owners or operators to prepare and submit response plans. These plans have to identify and ensure the availability of personnel and equipment required to remove (to the maximum extent practicable) a worst case spill and to mitigate or prevent a substantial threat of such a spill. The OSRO classification process was established in order to facilitate the preparation and review of these response plans.

The classification process represents standard guidelines that federal agencies and plan developers can use to evaluate an OSRO's ability to respond to and recover oil spills of various sizes. The OSRO classification process is strictly voluntary, and individual organizations are classified by geographic area and operating environment and by maximum most probable and worst case discharge. Classifications are based on response times, amounts and type of boom, recovery (skimmer) capabilities, temporary recovered liquid capabilities, and vessels. OSROs also have to demonstrate appropriate training programs and equipment preventive maintenance programs, and they may also provide information on dispersants, marine firefighting, and salvage.

The USCG Response Resource Inventory system maintains the OSRO listing. Alaska Chadux and ASRC Energy Services operate in both the Nome and Prudhoe Bay areas of the Western Alaska Captain of the Port, and another two organizations—Clean Harbors Environmental Services and Alaska Clean Seas—operate only within Prudhoe Bay. However, the operating environments within which they are capable of spill response vary widely. ASRC Energy Services operates in the nearshore, with a vessel capable of handling a maximum most probable discharge. Alaska Clean Seas operates in all areas, with facility and vessel responses up to Level 2 for offshore, nearshore, and ocean areas. State-approved oil discharge prevention and contingency plans in Alaska also require Primary Response Action Contractor (PRAC). A company applying for PRAC has to identify the region where their services will be made available, response personnel available, inventory of oil spill response equipment including dispersants and in situ burn capabilities, significant ancillary resources and equipment, and previous oil spill activities and history of compliance with state and federal environmental laws. There are currently eight PRACs registered for the Northwest Arctic and North Slope regions.

Oil Spill Response Equipment and Resources

U.S. Coast Guard

Command post locations for the Arctic are problematic due to the scarcity of infrastructure. Moreover, movement of response equipment to the scene of the spill will likely be difficult and time-consuming if not prestaged. The subregional contingency plans note that there is no government spill response equipment prestaged in the Arctic regions of Alaska, although there is industry equipment prestaged in the region. In any event, government equipment is meant as a supplement to the RP's efforts. The USCG maintains agreements with several spill response companies in Alaska to respond to an oil spill if the responsible party cannot be identified or cannot respond appropriately. Local, state, and industry spill response equipment is staged along the coast for responding to local spills related to exploratory drilling, oil fields, tank farms, and the Trans-Alaska Pipeline. Due to logistical restraints, this equipment does not provide capabilities for all possible spills—for instance, a transiting vessel. With the forecasted increase in shipping through the U.S. Arctic, prepositioning of additional equipment for long-range response capabilities to remote areas may need to be considered by the U.S. Coast Guard and/or the State of Alaska.

The USCG does not have permanently deployed assets in the North Slope area. The nearest cache of significant Coast Guard–owned equipment is in Anchorage. Other prepositioned sites exist in the southern areas of the state but are intended primarily for minor spill response or as a supplement to commercially provided equipment. Logistical support from Coast Guard aviation assets would be provided from Kodiak, which is approximately 1,600 km south of Barrow. In 2012, the USCG conducted the Arctic Shield exercise, which shifted resources from southern Alaska to the Arctic. Two helicopters were deployed to Barrow, national security vessels were brought into the Arctic during the summer months, and an oil spill exercise was conducted.[10] Arctic Shield is made possible by accepting tradeoffs in other Coast Guard missions, within the limits of the Coast Guard's budget (Marcario, 2013). A 2013 Arctic Shield exercise focused on the Bering Strait and western Alaska and included the use of a forward operation location in Kotzebue, in association with the National Guard.[11]

[10] See http://www.dhs.gov/news/2012/10/11/written-testimony-us-coast-guard-senate-commerce-science-and-transportation.

[11] See http://www.d17.uscgnews.com/go/doc/4007/1776075/; http://www.d17.uscgnews.com/go/doc/4007/1843778/; http://www.military.com/daily-news/2013/11/04/coast-guard-completes-arctic-shield-2013.html.

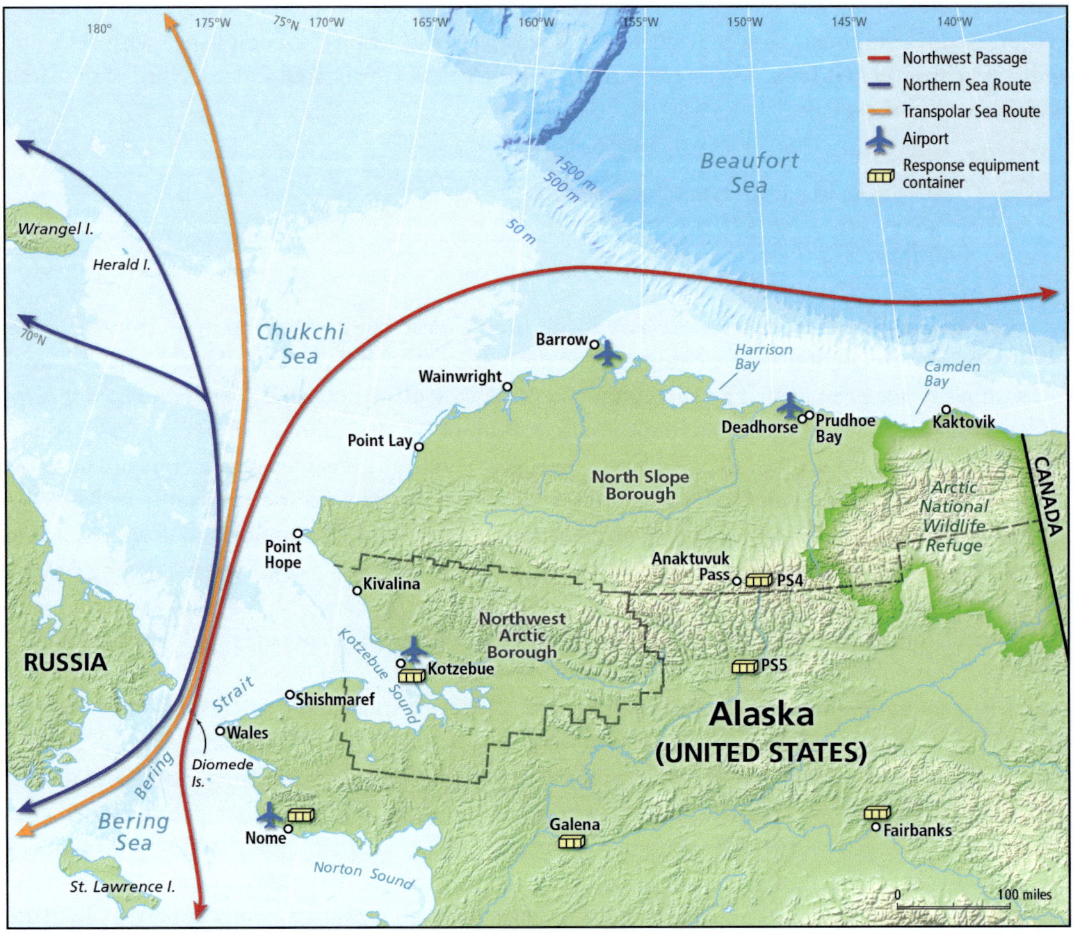

Figure 4.3 Arctic shipping routes, ADEC spill response equipment depots, and villages and towns with the capacity to land passenger jets in Alaska. Map area corresponds to the red box in Figure 1.1. All towns and villages have some spill equipment provided by their respective boroughs. Along the coast, Kotzebue, Nome, and Barrow also have spill response agreements with ADEC. While this map only shows passenger jet landing capabilities, almost all towns and villages have airstrips that can land C-130s.

State of Alaska

ADEC has several prepositioned response equipment depots in Alaska (Figure 4.3). The depots normally provide basic first aid capability in order to allow for timely response to a spill in a particular area without having to wait for outside resources to arrive. None of these packages are located on the North Slope.

ADEC has partnered with local communities to expand and improve oil spill response capabilities and readiness. ADEC has established more than 35 spill response agreements with boroughs and communities, which allow ADEC to request local assistance depending on the needs of a given incident. ADEC and communities that have signed agreements with them work together to improve preparedness and identify ways to enhance local capabilities through training and additional response equipment. If the local government assists in responding to local incidents, ADEC uses the agreement to allow for reimbursement of expenses and local training.

North Slope and Northwest Arctic Boroughs

The North Slope Borough (NSB) and Northwest Arctic Borough (NWAB) provide spill response equipment to villages in their areas of responsibility to support tank farm operations. While each of the villages has slightly different equipment available, it normally includes bales of sorbents, boom, skimmers, pumps, generators, personal protective gear, wildlife deterrence equipment, liner material, and support material. Shell recently provided additional spill response equipment for NSB.

Alaska Clean Seas

Alaska Clean Seas is a not-for-profit oil spill response cooperative. Its membership includes energy industry companies (oil and gas, pipelines) that plan to or currently work in North Slope oil and gas exploration, development, production, or pipeline transport activities. Its area of operation is Alaska's North Slope, the Outer Continental Shelf off the coast of the State of Alaska, and the Trans-Alaska Pipeline from Pump Station 1 to milepost 167. Equipment is owned by Alaska Clean Seas and member companies and is located in Prudhoe Bay. Boats in the inventory were designed to respond to incidents in the immediate area and do not have berthing or mess facilities; instead, they are designed to be supported from shore-based facilities. Their equipment includes boom, several different types of vessels, pumps, hoses, skimmers, storage tanks, ISB equipment, and support materials (ACS, 2012).

Alaska Chadux Corporation

Alaska Chadux Corporation is a member-funded, 501(c)(4) not-for-profit oil spill response organization. It is headquartered in Anchorage, Alaska, and provides resources for containing, controlling, and cleaning up petroleum spills for its member companies. Its headquarters and warehouse in Anchorage have resources packaged for rapid transport by land, water, or air. Alaska Chadux Corporation also has equipment at 12 strategically located hubs, including Barrow and Nome. The equipment staged in Barrow is minimal, consisting of boom, absorbent pads and sweep, and wildlife kits.

ASRC Energy Services Response Operations

ASRC Energy Services Response Operations, LLC (AES) was formed in 2006 to build a response capability to support Shell's offshore exploratory drilling activities in the Chukchi and Beaufort Seas. With the exception of some equipment in Wainwright, their equipment is located on site during the open water season only. Some of the equipment is available through Shell charter and not directly owned by AES. Their equipment includes boom, response vessels, skimmers, pumps, storage, dispersant, and generators.

There is a significant amount of spill response equipment located in the North Slope, especially when Shell and AES equipment are deployed to support exploratory drilling operations. The majority of the equipment is located in Prudhoe Bay and in the immediate vicinity of oil and gas exploration activities. However, other shoreside facilities to support spill response operations are limited.

U.S. Facilities for Arctic Spill Response Research and Training

Ohmsett

Ohmsett is the National Oil Spill Response Research and Renewable Energy Test Facility. It is located in Leonardo, New Jersey, and is maintained and operated by BSEE. At over 200 m long, the Ohmsett facility provides one of the largest outdoor saltwater wave/tow tank facilities and has been used for performance testing of a variety of oil spill response options, including mechanical equipment and dispersants under a wide range of environmental conditions and water temperatures. The facility also provides for oil spill research and development of new technologies, such as those that were evaluated during the Wendy Schmidt Oil Spill X Challenge. In addition to this, Ohmsett offers spill response training to personnel from government agencies, industry, and the private sector. They hosted an "Ice Month" in spring 2013 to test mechanical response equipment in broken ice and Arctic temperature conditions and conducted cold-water dispersant effectiveness tests in early 2014 (Ohmsett, 2013, 2014). There are also plans to test Arctic-specific mechanical recovery techniques in 2015 and 2016.

U.S. Army Cold Regions Research and Engineering Laboratory

The U.S. Army Cold Regions Research and Engineering Laboratory (CRREL) is a research facility within the U.S. Army Corps of Engineers' Engineer Research and Development Center that maintains a particular focus on cold regions. Its mission is to develop solutions to current and emerging problems by advancing science and engineering in a number of technical areas, including biogeochemical processes, infrastructure, environmental fate and transport geochemistry, and hydrology. A key feature of CRREL's cold weather-related facilities is the Ice Engineering Test Basin, which has been used to support a variety of research and personnel training activities, including those of the U.S. Navy, USCG, and industry. The tank is 37 m in length and may be cooled as low as -24°C. CRREL facilities have also been used by groups, including Alaska Clean Seas, to train personnel on oil detection and response in icy conditions.

OTHER INFRASTRUCTURE

Commercial infrastructure is either limited or absent in the U.S. Arctic. Oilfields around Prudhoe Bay host support service contractors and their equipment. In the event of a SONS and the necessity for rapid deployment of large numbers of responders, passenger jet service (737-scale) is available at Nome, Kotzebue, Barrow, and Deadhorse (Figure 4.3). Smaller aircraft service (19-passenger turboprop) can access nearly all of the approximately 30 coastal communities and other developments (e.g., Red Dog Mine, De Long Mountains Terminal) from Nome to the Canadian border. Almost all of the airstrips can be accessed by C-130 and smaller cargo aircraft if needed for rapid deployment of spill response equipment. Multiple heavy lift aircraft would be needed to bring in capping stack equipment.

Spill responders and other personnel would find a severe shortage of housing, fresh water, food and catering, sewage handling and garbage removal facilities, communications infrastructure, ability to handle heavy equipment, supplies, and hospitals and medical support. Large numbers of response workers also represent an increased risk of accidents and injuries. There are also limited bandwidth and communications capabilities. A single fiber optic cable connects the existing oil fields and there are currently no cables to northwestern Alaska, although a hybrid of fiber optic and microwave repeater towers are planned for the Northwest Arctic Borough. Increased bandwidth capacity is needed to share data and information in the event of an oil spill.

Moreover, recovered oil and oily debris must be collected and disposed of in predesignated locations, or the means to transport the material to some approved location outside of the local area is needed. Given the limited highway infrastructure, planners will inevitably look to aviation and seaborne support for all of these needs. There are no deepwater ports in the two boroughs. Nome has a shallow water port with docks, while other villages have shallow embayments (0-20 ft) without support facilities. The distance from Dutch Harbor, the closest full-service port, to the Shell drilling site in the Chukchi Sea, for example, is approximately 1,600 km. Sea-based support will be limited in its ability to work very close to the shore, due to shallow waters in much of the region, so a contingent of shallow water craft is needed for nearshore operations. Most of this can be contracted commercially, provided through government or military sources (if available), or provided on station by the operator and ready for immediate use. This latter approach was followed by Shell during its 2012 season. Absent this approach, the time delay in bringing adequate capabilities to the scene could be significant.

Infrastructure Assets for Environmental Monitoring

Real-time sea ice and water-column measurements of temperature, salinity, and velocity are needed to understand ocean stratification and surface and subsurface circulation for the fate of spilled oil and for better numerical models of oil spill pathways. Some current observing capabilities are described in this section.

University of Alaska Fairbanks (UAF) infrastructure and instrument assets include shore-based, high-frequency radar (HFR) systems that map surface ocean currents (Weingartner et al., 2013a). Installation sites in Barrow, Wainwright, Point Lay, and Cape Simpson allow for data collection within 200 km of the coastline over an area of approximately 30,000 km^2. Data are fed to the National HF Radar Network[12] and are also available from UAF.[13] Current vectors are calculated at 6-km spatial resolution and represent velocity in the upper 1 to 2 m of the water column. The systems can only operate during the open water season (approximately July to October), and waves must be present to measure currents. There can be spatial data gaps close to shore, due to the geometry of the coastline. Fully automated solar–wind energy hybrid power modules have been developed to provide power to the HFRs, meteorological sensors, and satellite-based communications. In the case of an oil spill, additional HFR systems could be installed in about three days at remote sites, with plans to create a more portable system that could be deployed in one day. An expansion of the HFR array is being considered, where additional systems could be maintained by local communities (Thomas Weingartner, UAF, personal communication).

Three UAF autonomous underwater vehicles (AUVs) are operating over most of the northeast Chukchi Sea[14] and can sample an area for up to four months (see Weingartner et al., 2013b). These gliders presently cannot operate under sea ice cover, as developments are needed for under-ice navigation and through-ice communications. Pilot missions under sea ice will be made in 2014, as well as research missions as part of the Office of Naval Research Emerging Dynamics of the Marginal Ice Zone initiative and the European Union Arctic Climate Change, Economy and Society project. In addition to standard temperature and salinity measurements, AUVs can be equipped with sensors to measure colored dissolved organic matter and chlorophyll, to map ice topography by multibeam sonar, and to compute water velocity.

Instrument systems towed from ships are also used to measure temperature, salinity, and bio-optical properties at high horizontal resolution in the water column and can be used when conditions do not permit the use of AUVs—during strong ocean currents, for example. Additional applications of AUVs and towed vehicles include autonomous mapping of plumes (e.g., sediment, oil) and ice features (e.g., draft, keel depths, bottom scouring), mapping and imagery of hydrographic and bottom features, and incorporating water sampling with biochemical optical measurements for chemical laboratory analysis (e.g., Wadhams et al., 2006; Wilkinson et al., 2007). To supplement these data, moorings anchored to the seafloor can collect water-column measurements, as well as information on overlying sea ice drift and thickness, in one location over seasonal or longer timescales.[15]

Ice-tethered profilers are automated, drifting ocean profiling instruments that are deployed in drifting sea ice or in open water conditions to sample physical and some biological parameters in the upper water column (to 750 m depth) during all seasons, providing essential data on vertical ocean stratification for assimilation into numerical forecast models (Krishfield et al., 2008; Toole et al., 2010). They transmit ocean data in near real-time from surface buoys, with horizontal resolution on the order of 1 km and vertical resolution on the order of 25 cm. At present, they are deployed in central Arctic Ocean basins, but future ice-tethered profilers designed for shallower profiling depths could provide these data for the shallower Chukchi Sea and boundaries of the Beaufort Sea.

[12] See http://cordc.ucsd.edu/projects/mapping/.

[13] See http://dm.sfos.uaf.edu/chukchi-beaufort/index.php.

[14] See http://www.ims.uaf.edu/artlab/instruments/gliders.php.

[15] See, for example, https://www.whoi.edu/beaufortgyre/.

Real-Time Traffic Monitoring

There is no comprehensive system for real-time traffic monitoring. The capability that comes closest to achieving maritime awareness is that established by the Alaska Marine Exchange, which uses Automatic Identification System (AIS) receivers to track vessels in the strait and along a large percentage of Alaska coastal areas. However, its coverage is not 100%, with significant gaps in northwestern Alaska and along the North Slope (Pt. Barrow and Prudhoe Bay being notable exceptions). Satellite-based AIS reception typically involves a time delay to coincide with overhead satellite passes, versus a continuous real-time display available with terrestrial-based receivers. Antenna placement on vessels may also influence signal receptivity by satellites, since AIS was designed as a line-of-sight terrestrial system. Moreover, the Coast Guard does not have its own AIS receiver system in place for this region and therefore maintains a close relationship with the Marine Exchange. The data from the Marine Exchange supplements its own on-scene presence through seasonal deployments of aircraft and cutters. The Coast Guard is currently evaluating traffic through the Bering Strait in order to determine if an internationally recognized traffic separation scheme is warranted (RADM Thomas Ostebo and 17th Coast Guard District personnel, personal communication, February 4, 2013, and March 18, 2014).

Incident Command Centers and Local Inclusion

The subregional contingency plans recognize the significant challenges for the RP and government to identify suitable Arctic command posts. Based on discussions with federal and state responders, it is very likely that a primary Incident Command Post would be located in Anchorage, which has sufficient infrastructure to accommodate the response effort. This has significant drawbacks in that it would be far removed from the spill scene in an Arctic event. For this reason, responders indicate that it would most likely be necessary to have a forward operating base to exercise tactical control over the effort. This is a reasonable expectation, and would greatly enhance the ability to absorb local expertise into the response effort. Absent a forward command post, local input would be more difficult to access, to the detriment of the overall response effort. An alternative would be to bring local experts to the primary Incident Command Post to enable consultation on a real-time basis, although this would be less desirable in that direct observation of on-scene conditions would be lacking. Fundamentally, the absence of sufficient inclusion of local on-site expertise can easily lead to ill-considered response decisions. Informal discussions with tribal leaders have indicated concerns that they will not be consulted in real time, and that critical decisions will be made in remote locations, using computer modeling, not validated by local knowledge and observations.

Workshops held in March 2013 in Alaska brought together indigenous experts and scientists to discuss ice and ocean conditions in the Beaufort and Chukchi region with respect to oil spill and other incident response (Johnson et al., 2013). The significant value of the collective knowledge of the ice-ocean system was demonstrated, in particular the understanding that local knowledge is essential for response planning. Recommendations of the collaborative study included identification of emergency shoreline staging locations and the ability to provide real-time tracking of very large floes.

INTERFACING WITH LOCAL GOVERNMENTS

The NSB is the nation's largest municipality in terms of land area. It is a relatively strong home-rule government with a significant physical presence in each of its eight remote communities. In the smaller communities especially, the NSB performs nearly all of the physical infrastructure support, from road and runway maintenance to power plant and water/sewer operations. During storms, floods, or other natural emergencies in or near one of its villages, the NSB has used its heavy equipment fleet to respond. A fleet is kept in operational condition in each of the villages. When compared to the NSB, the NWAB has less of a dominant presence in each of its respective communities. In NWAB villages, physical assets such as heavy equipment, bulk fuel tank farms, and power plants are more likely to be owned and operated by organizations such as the local tribes, school district, or private enterprise. The Alaska Native Corporations also have varying amounts of physical assets that could be of use during an oil spill response. In addition, some of the Alaska Native Corporations are employed by the oil and gas industry for science and data gathering, logistical support, oil spill response, and other support services.

The NSB has a relatively well-developed and staffed Local Emergency Response Planning and Coordination (LERPC) team. Because of its lack of a large operational base, the NWAB does not have as strong a presence in disaster response scenarios, but they are interested in strengthening this capability. The NSB LERPC team has developed a set of scenarios (e.g., catastrophic tank farm failure, aircraft crash, and storm and flood events) that are outlined, updated, and practiced in some form using the ICS structure. For recent emergencies, the NSB, following its LERPC plan, has fielded ICS teams to the nearest affected village. The team typically consists of high-level staff from the NSB Mayor's Office, Planning Department, Public Safety, Fire, Search and Rescue, and Municipal Services, with public information officers and others taking part. In the 2012 gas blowout of an onshore exploration well near Nuiqsut, the NSB dispatched its team, which was incorporated into state and industry ICS teams to form an integrated multiagency response and command structure. However, there is little connection between the NSB ICS team and industry-supported drills and exercises, which tend to be self-contained. The LERPC team operates more or less independently of the Unified Command structure. Some of this may be attributed to the physical distance between Incident Command Posts and the community-based capabilities. There may also be a technological bias, which places greater emphasis on such tools as computer models and aerial imagery than on subjective experiential information.

In all Alaskan native regions, individual communities look toward elders and other community members with the most relevant experience and environmental and/or traditional knowledge. While they may have no formal role, elders and other experts often direct or assist in efforts at the highest local level. Each Arctic community also has a strong network of volunteers that are typically active in emergency response, search and rescue, and related efforts. Volunteer teams in the North Slope are coordinated with the NSB-supported Emergency Response Planning and Coordination teams.

A regional citizens advisory council, similar to those established in Prince William Sound and Cook Inlet, Alaska (CIRCAC, 2012; Prince William Sound Regional Citizens Advisory Council, 2013), has been proposed by several organizations (National Commission on the BP Deepwater Horizon Oil Spill and Offshore Drilling, 2011). This entity could engage in the planning process for Arctic exploration, and protect the food supply, health, and culture of Alaska natives. Neither Congress nor the Administration has implemented this recommendation, although it is included in proposed legislation (OSCA, 2013).

HIGH-RELIABILITY ORGANIZATIONS

High-reliability organizations are systems of organizations that deliver high levels of organizational and mission performance in challenging operational settings (Roberts, 1990). These organizations, by nature or design, cannot or must not fail, because the consequences of failure are usually catastrophic. Examples include nuclear power plants, naval command and control organizations, commercial airline operations, chemical process plants, distributed transportation systems, and various facets of health care and medical systems (Grabowski and Roberts, 1999, 2011). Arctic oil spill response operations also have significant mission, safety, and reliability mandates that must be met in the face of events of varying scales and sources, substantial geographic distances, and extreme environmental conditions of visibility, daylight, wind, weather, ice, and snow.

Large-scale, distributed oil spill response organizations in the Arctic are also in many cases virtual organizations—electronically linked, distributed systems of organizations tied together by a common mission and common objectives, whose members may or may not know each other or work together face-to-face (Davidow and Malone, 1992; Grabowski et al., 1997, 2007; Grabowski and Roberts, 1999). Virtual organizations with high-reliability mandates are known as high-reliability virtual organizations—distributed, large-scale systems of organizations, whose members are electronically linked and who share a common value chain and a mandate for mission and safety reliability (Grabowski and Roberts, 1997, 1999; Grabowski et al., 2007; Grabowski and Roberts, 2011).

Preparing for Large-Scale, Distributed Arctic Oil Spill Response

The ability to provide flexible and scalable organizational structures in response to environmental demands is essential for the success of highly reliable virtual organizations, as they must vary their organizational structures from low- to high-tempo operations as environmental conditions change (Roberts, 1990; Harrald et al., 1992). This was demonstrated during the *Kulluk* operational response (Box 4.1).

Inclusive Decision Making

Highly reliable organizations are known for empowering local decision makers with authority and responsibility for their areas of operation (Roberts, 1990; Grabowski and Roberts, 1996). Inclusion activities permit the development of organizational and cultural trust among all participants (Franken and Thomsett, 2013). The challenge for Arctic oil spill response is to work inclusive decision making into the structure so that the appropriate stakeholder input is included from the outset (Pew Charitable Trust, 2013). Efforts at outreach and inclusivity in Arctic oil spill response preparation and planning have been made by Shell, ConocoPhillips, and Statoil and by state and

Box 4.1
Organizational Lessons Learned from the Grounding of the *Kulluk*

As mentioned in Box 1.1, Shell's conical drilling unit *Kulluk* grounded off Sitkalidak Island on December 27, 2012, after becoming separated from its tow connection. The Incident Command System (ICS) formed the basis for the *Kulluk* response organizational structures. The *Kulluk* ICS incident response was expanded to a Unified Command with broadened participation in the Command Center, including local experts, stakeholders, and the media. The initial *Kulluk* response organization was small, but it grew to well over 500 participants over the course of the response in January 2013. As the event resolved, the size of the response organization shrank. Flexible and scalable organizational structures during the *Kulluk* response were intentional results of several years of cross-training, simulations and training exercises that predated the actual event, as well as the result of pairing operational personnel with their counterparts from different functional areas during oil spill events. The ease of organizational growth and adaptation seen during the *Kulluk* response was attributed by a senior observer to system members' familiarity with the Unified Command System (UCS) and ICS developed over frequent (thrice-yearly) oil spill response drills and in other types of emergency response activities for which the participants were trained: for hurricanes, tsunamis, and even for Y2K events. Both public and private information systems were utilized during the *Kulluk* event, with public information available to all participants, but private, company-sensitive information only available to a smaller group of decision makers. To provide updates, situational reports, and information to the interested public and the media, many organizations, including federal and state participants, utilized social networking during the *Kulluk* event. However, social media communication about the event was not integrated in a common platform or display. Traditional media outlets and information sources during the *Kulluk* event suffered the same lack of integration.

federal agencies (UNOLS, 2013; Young, 2013). In one recent drill, for instance, the Lummi tribe played a substantial role in the Incident Management Team command of a drill in Washington State (Tom Coolbaugh, personal communication, 2013). In addition, an important focus in USCG Arctic Shield 2012 activities was the development of partnerships with native and local stakeholders in order to build the social capital that is critical during emergency and disaster response activities (Young, 2013). During the *Kulluk* event, local participants joined the Incident Management Team command structure after several days had passed, when it became clearer who would be affected. However, the managing of stakeholder participants has led to recommendations for changes in organizational structures and information availability after the event (ARRT, 2013).

Common Operating Picture

The lack of a common and standard information technology architecture, processes, and software baselines proved highly problematic in the early stages of the *Deepwater Horizon* oil spill response (Konrad and Shroder, 2011; USCG, 2011). Public confidence in a response effort can be undermined

Scenario 4
Tug and Barge Accident

A barge that operates for part of the year in the Beaufort Sea is being towed out of the Arctic in late October, a time of year when ice formation in the Arctic is under way. It carries nearly 150,000 gallons of diesel fuel and about 10,000 gallons of jet fuel. A severe storm separates the tug from the barge and, despite attempts by the USCG and other dispatched vessels, weather and ice conditions preclude rescuers from being able to reattach towlines. Attempts continue for several days, but ultimately the barge grounds.

Main Considerations
The main issue will be reaching the scene of the incident, which can be hampered by poor weather, low visibility, and possible ice. Secondary considerations include the potential for an oil spill and contamination of the shoreline in remote areas, which will also encounter challenging logistics. Cleanup could extend over the winter months. Also, liability limits will be reached relatively quickly because limits for barges under OPA 90 are less than for ships.

Standard Response
Responsible party assumes leadership role for cleanup, an oil spill removal organization (OSRO) is identified (depending on the tow), and salvage capabilities and response are identified and activated.

Response Needs
Adequate training and exercises for combined or complex events (salvage/personnel rescue/oil spill response) scenarios, resources for potential shoreline contamination cleanup, training and exercises for local response personnel working with Unified Command personnel in winter, and logistics in remote areas are needed.

if consistent, authoritative, and accurate information cannot be provided (Pew Charitable Trust, 2013). A flexible and open common operational platform for gathering, displaying, and sharing data can ease the time spent in responding to queries and resolving data conflicts. Shared and standard information technology architectures and technology are particularly important in highly reliable virtual organizations, as participants are geographically scattered, with varying degrees of access to and understanding of the on-the-ground operational situation. Distributed decision makers linked by large-scale electronic tethers need to develop a shared, common operational understanding and shared mental models of the response requirements and effort as it unfolds. A common operational picture of an unfolding event thus offers a number of advantages to large-scale, distributed Arctic oil spill response. Integrated systems that meld public and private information can provide important, synthesized input to decision makers and can support the development of shared mental models of unfolding operational scenarios, although protecting and disaggregating different participants' proprietary or sensitive information is a recurrent challenge in distributed, virtual organizations. There is a need for both traditional and nontraditional technology elements (e.g., social media, location-based services, cloud infrastructure support, mobile applications) and for adherence to international standards for software development, support, and security.

In its Arctic Strategy, the Coast Guard has proposed an Arctic Fusion Center to promote cooperation and coordination; to leverage joint, interagency, and international capabilities; and to enable sustainable development and environmental protection (USCG, 2013b). Protocols such as the European Union's Common Emergency Communication and Information System and tools such as the Virtual On-Site Operations Coordination Center may provide mechanisms that ease integration and data sharing among nations, agencies, organizations, indigenous people, stakeholders and the public (Arctic Council, 2012). However, additional international protocols and tribal cooperation and agreements will be needed, beyond the resources and capabilities of a single government, entity, or agency, in order to bring together the needed data, information, integration, and dissemination capabilities needed.

CHAPTER CONCLUSIONS AND RECOMMENDATIONS

Conclusion: Marine activities in U.S. Arctic waters are increasing without a commensurate increase in the operations, logistics, and infrastructure needed to conduct these activities safely. U.S. support for Arctic missions, including oil spill response, requires significant investment in infrastructure and capabilities, such as transportation, communication, energy and fuel, electricity, housing/berthing, navigational aids, charting, port access, ice forecasting, ship repair, and salvage.

Conclusion: The United States has bilateral agreements with Canada and Russia regarding oil spill response. Formal contingency planning and exercises with Canada have enabled both the United States and Canada to refine procedures and legal requirements for cross-border movement of technical experts and equipment in the event of an emergency. Exercising the bilateral agreement with Russia will more fully enable both countries to address practical issues that could arise in an actual spill. An active exercise program with Russia, similar to that with Canada, could identify problems and resolve them in advance.

Conclusion: The Russian Federation's commitment to the economic development of the Northern Sea Route has expanded the volume of large-vessel traffic through the Bering Strait, with greater possibility of a major vessel accident and implications for environmental impacts in U.S. waters. The resolution of anticipated response problems, such as communications between command centers, coordinated planning, transboundary movement of people and equipment, and identification of translators, needs to be accomplished in advance of an actual event.

Recommendation: The USCG should expand its bilateral agreement with Russia to include Arctic spill scenarios and conduct regularly scheduled exercises to establish joint responses under Arctic conditions and should build on existing bilateral agreements with Russia and Canada to develop and exercise a joint contingency plan.

Conclusion: Vessel traffic is not actively managed in the Bering Strait or in the U.S. Arctic, nor is there a comprehensive system for real-time traffic monitoring. The lack of a U.S. vessel traffic monitoring system for the Arctic creates significant vulnerability for U.S. Arctic missions, including oil spill response, and creates undue reliance on private industry and foreign national systems. Private AIS receivers are used to track vessels in the Bering Strait and along a large part of Alaska coastal areas, but there are significant gaps in coverage. Consequently, there are numerous regional "blind spots" where an early indication of elevated risks may not be apparent to officials on shore.

Recommendation: The USCG should expedite its evaluation of traffic through the Bering Strait to determine if vessel traffic monitoring systems, including an internationally recognized traffic separation scheme, are warranted. If so, this should be coordinated with Russia. The USCG should also consider obtaining broader satellite monitoring of AIS signals in the Arctic through government means or from private providers.

Conclusion: The Oil Spill Liability Trust Fund may prove insufficient to cover the sociological as well as economic damages of an affected community. A structure other than the National Contingency Plan may be needed to deal with broader social impacts resulting from a significant oil spill. One approach could be to amend the Stafford Act and the Oil Pollution Act of 1990 in order to enable funding for a "whole government" method if it is determined that the Oil Spill Liability Trust Fund would be insufficient for the purpose. In this case, the National Response Team would need to consider how a response conducted under the National Contingency Plan could blend with a Stafford Act response structure, should the need arise.

Conclusion: The USCG has a low level of presence in the Arctic, especially during the winter. Coast Guard personnel, equipment, transportation, communication, navigation, and safety resources needed for oil spill response are not adequate for overseeing oil spill response in the Arctic. The Coast Guard's efforts to support Arctic oil spill planning and response in the absence of a dedicated and adequate budget are admirable but not sustainable.

Conclusion: Prepositioning a suite of response equipment throughout the Arctic, including in situ burn and dispersant capability, would provide immediate access to all oil spill countermeasures. In remote areas, prepositioning and maintenance of fuel caches could be critical to extend aircraft and helicopter range. Storage and maintenance needs for prepositioned caches will need to be considered, as will the resources to support them in the long term.

Recommendation: As oil and gas, shipping, and tourism activities increase, the USCG will need an enhanced presence and performance capacity in the Arctic, including area-specific training, icebreaking capability, improved availability of vessels for responding to oil spills or other emergency situations, and aircraft and helicopter support facilities for the open water season and eventually year round. Furthermore, Arctic assignments for trained and experienced personnel and tribal liaisons should be of longer duration, to take full advantage

of their skills. Sustained funding will be needed to increase the USCG presence in the Arctic and to strengthen and expand its ongoing Arctic oil spill research programs.

Conclusion: The absence of infrastructure in the U.S. Arctic would be a significant liability in the event of a large oil spill. It limits the ability to conduct routine operations and maintenance, engage local communities, and develop meaningful area familiarity. There is presently no funding mechanism to provide for development, deployment, and maintenance of temporary and permanent infrastructure. One approach to provide a funding mechanism for infrastructure development and oil spill response operations would be to enable a public-private-municipal partnership to receive a percentage of lease sale revenues, rents, bonuses, or royalty payments that are currently deposited in the federal treasury.

Conclusion: Effective oil spill response requires improved communication bandwidth and networks; transportation systems; environmental and traffic monitoring systems; energy and fuel systems; personnel, berthing, housing, waste and medical support facilities; as well as civil infrastructure development to provide improved port and air access to remote locations using extended supply chains and an increased capacity to handle equipment, supplies, support, and personnel. Strategic development of multiuse facilities (e,g., schools, community buildings, and gymnasiums) would enable them to be used as response control centers. Human and organizational infrastructure improvements are also required to improve international and tribal partnerships so as to leverage scientific and traditional knowledge and best practices.

Recommendation: Infrastructure to support oil spill response should be enhanced in the North Slope Borough, Northwest Arctic Borough, and communities along the Bering Strait,[16] with marine facilities for addressing response operations. The scope, scale, and location of infrastructure needs should be determined through structured decision processes, studies, and risk assessments.

Conclusion: Local communities possess in-depth knowledge of ice conditions, ocean currents and potentially affected marine life in areas that could be affected by oil spills. Failure to include local knowledge during planning and response may risk missing significant environmental information, yet there appear to have been only modest efforts to integrate local knowledge into formal incident command-based responses. Developing and maintaining trained village response teams would integrate local knowledge and utilize existing human resources for effective oil spill response.

Recommendation: The USCG and Alaska Department of Environmental Conservation should undertake the development of an oil spill training program for local entities so as to develop trained response teams in local villages. Industry should continue to participate in

[16] The wording of this recommendation was edited after release of the prepublication to explicitly include communities along the Bering Strait.

local training initiatives. Local officials and trained village response teams should be included in the coordinated decision making and command process during a response event. Input from community experts should be actively solicited for inclusion in response planning and considered in conjunction with data derived from other sources. The Coast Guard should set this as an exercise objective in all government-led oil spill response exercises in the Arctic and should set the expectation that industry-led exercises will do the same.

Conclusion: An end-to-end system that integrates Arctic data in support of preparedness, response, and restoration and rehabilitation for oil spills is needed. To achieve this, development of international standards for Arctic data collection, sharing, and integration is required. Interoperability between systems would reduce duplication and increase availability of a common operational picture shared among all participants and stakeholders. A standard architecture that synthesizes multiscale, heterogeneous data from multiple sources, including traditional and local knowledge, is an important next step to address Arctic data challenges. Because a number of U.S. agencies have responsibility for these data and system needs, the Interagency Arctic Research Policy Committee could act as a coordinating body.

Conclusion: Flexible and scalable organization is important to develop an effective Arctic oil spill response. This can be achieved through drills, case studies, simulation, and organizational learning. To build the systemwide capacity to respond to large-scale, distributed Arctic oil spill response, sustained long-term training and continued resource investments are required. Inclusive and trustful communications, relationship building, and decision making, clear accountability, and ongoing assessment and efforts for improvement are also necessary.

Recommendation: Relevant federal, state, and municipal organizations (such as USCG, NOAA, BSEE, the Bureau of Ocean Energy Management, Alaska Department of Environmental Conservation, Alaska Department of Natural Resources, U.S. Fish and Wildlife Service, Alaska Fish and Game, North Slope Borough, and Northwest Arctic Borough), local experts, industry, and academia should undertake regularly scheduled oil spill exercises designed to test and evaluate the flexible and scalable organizational structures needed for highly reliable Arctic oil spill response.

5 Strategies for Response and Mitigation

PRE-SPILL STRATEGIES FOR OIL AND GAS EXPLORATION AND PRODUCTION

Essentially all pre-spill strategies, whether in the Arctic or elsewhere, emphasize prevention as the key oil spill–related activity. Prevention is based on understanding the science and technologies associated with oil exploration and production and applying that understanding to the specific environment in which any activity is taking place (USGAO, 2012), with the goal of managing risk throughout an operation. With a high level of knowledge, many potentially high-risk activities can be anticipated and may be mitigated. For example, when oil wells are properly designed for the range of anticipated risk, established procedures are followed, equipment is properly inspected and maintained, and training is provided, well control events are likely to be managed more safely (ExxonMobil, 2013). When accidents occur, the causes and outcomes can be analyzed so that new information can be incorporated in future plans.

In general, a tiered response system (Figure 5.1) represents the standard worldwide approach to evaluating risks and resource needs (e.g., ExxonMobil, 2013). Tier 1 is exemplified by a small oil spill that can be handled quickly at the local level. If weather, such as high winds or seas, becomes a factor, this could be elevated to a Tier 2 response, and rapid response and control may not be possible for a variety of reasons, including concerns for responders' safety. As previously mentioned in Chapter 4, regional resources could also be limited in Arctic locations. In the case of a large spill, a Tier 3 classification could occur quite rapidly to make sure that needed resources are put into play as quickly as possible. The potential for resource limitations in a specific region—for instance, the Arctic—is identified in contingency plans.

The oil and gas companies that work in the Arctic belong to the major Tier 3 oil spill removal organizations (OSROs), as well as to more regionally focused OSROs that have significant cold-region response capability and experience. A common strategy of Arctic and temperate spill response involves making sure that all possible countermeasure tools are available for use as conditions permit (e.g., Owens et al., 1998; Dickins and Buist, 1999; Brekne et al., 2004; Cater, 2010).

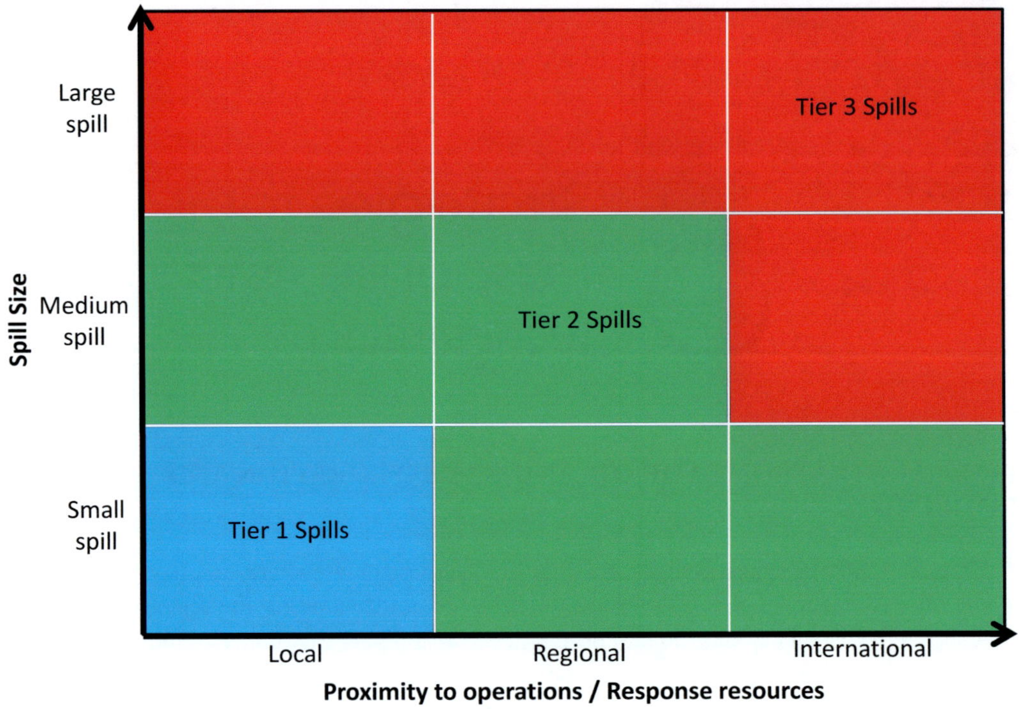

Figure 5.1 Example of a tiered response system. SOURCE: ExxonMobil (2013).

INDUSTRY ACTIVITIES POST–*DEEPWATER HORIZON*

In addition to ongoing development and review of specific oil spill prevention and response plans in support of exploration operations, a number of response-related activities took place immediately following the *Deepwater Horizon* oil spill. Industry task forces were created to review options associated with containing oil spills at their source, especially subsea releases. While the focus has been on the Gulf of Mexico, well containment technologies could be used in other exploration areas including the Arctic (USGAO, 2012).

The U.S.-based Joint Industry Task Force[1] (JITF) created task forces to examine offshore operating procedures, offshore equipment, subsea well control and containment, and spill preparedness and response.[2] The International Association of Oil & Gas Producers created a Global Industry Response Group[3] (GIRG) to ensure that the lessons learned from recent wellhead blowouts and

[1] Members of the JITF include the American Petroleum Institute, International Association of Drilling Contractors, Independent Petroleum Association of America, National Ocean Industries Association, and United States Oil and Gas Association.

[2] See http://www.spillprevention.org/jitf-overview.html.

[3] See http://www.ogp.org.uk/publications/management-committee/capping-and-containment-global-industry-response-group-recommendations-japanese-translation/.

Scenario 5
Break in Subsea Pipeline from Nearshore Production

A break in a subsea pipeline at locations such as North Star, Ooguruk, or Nikaitchuq would lead to oil spilling into the nearshore environment and possibly moving out into the offshore region. An alternative scenario is a chronic, long-term-duration pipeline leak below detectable limits.

Main Considerations
The failure of a pipeline could potentially lead to a large-scale oil spill, with the possibility of oil spilled in the nearshore to move offshore or onshore. However, there could also be only limited oil spill release due to automatic valve closures. There may be long-term human and ecosystem risks posed by chronic low-level leaks.

Standard Response
Responsible party assumes leadership role for cleanup; Unified Command is activated.

Response Needs
Environmental benchmark and monitoring data, a multiuse Arctic observing network, high-resolution aerial mapping, and adequate training and exercises for local responders working with the Unified Command and pipeline personnel, possibly over winter or in adverse weather, are needed.

other accidents are applied around the world. The Subsea Well Response Project[4] was organized around four tasks: designing a subsea well-capping toolbox, designing hardware for subsea dispersant injection, assessing the need for and feasibility of a containment system suitable for international use, and evaluating potential approaches for equipment deployment.

Each group made recommendations, mostly focused on well containment and intervention; oil collection, processing, and storage; and relief wells. Several multiyear projects have been initiated in efforts such as good practice guides for spill response, research programs to enhance spill response options, and outreach activities to assist decision makers, communities, and researchers to better understand oil spill prevention, preparedness, and response (Potter et al., 2012). GIRG's recommendations centered around prevention as a way to achieve the most effective outcomes.[5]

In addition to these industry-wide efforts, there has been a focus on creating new technologies for well containment and spill response. Several organizations can provide access to significant response tools as well as knowledge and skills to build capacity. The members of the Marine Well Containment Company[6] drilled approximately 70% of the U.S. Gulf of Mexico deepwater wells from 2007 to 2009. Helix Well Containment Group[7] is a consortium of 24 operators that represent

[4] See http://subseawellresponse.com/.
[5] See http://www.ogp.org.uk/downloads/GirgBrochure.pdf.
[6] See http://marinewellcontainment.com/.
[7] See http://www.hwcg.org/.

approximately 80% of the deepwater operators in the Gulf of Mexico. It was created around the well containment capabilities used in the *Deepwater Horizon* response. The Subsea Well Intervention Service[8] has an integrated intervention system that includes well capping and dispersant equipment that can be deployed internationally in the event of a subsea well control incident. The United Kingdom–based Oil Spill Prevention and Response Advisory Group[9] focuses on technical issues, including first response for protection of personnel, oil spill response capability and remediation, and regulations and response mechanisms. They have a well-capping device that enhances the United Kingdom's capability to respond to a major oil spill.

Independent oil companies also have capping stacks that can be designed for specific areas of deployment—for example, extreme cold and ice for Arctic environments—or may be transportable to other areas by vessel or large aircraft. While the ability to transport equipment by air enhances response capabilities outside of the Gulf of Mexico, it may not be very feasible. The capping stacks would have to be disassembled (because they are so large), loaded on multiple aircraft, reassembled on the other end, and tested to ensure integrity. Transport by vessel may be more likely, since there are multiple capping stack locations around the world. In general, these organizations have significant experience with deepwater drilling operations and response needs.

RESPONSE STRATEGIES FOR IMPACTED WILDLIFE

In the event of an oil spill, wildlife, including birds and marine mammals, could be impacted. Impacts could be due to contact with oil or disturbance from spill response activities. Annex G of the Unified Plan guides any responses to wildlife that are threatened or impacted by a spill event in Alaska (ADEC Spill Prevention and Response, 2002a). Wildlife response strategies are categorized as primary, secondary, and tertiary. The primary strategy is to control oil release and spread at the source through mechanical containment and recovery, protective booming, in situ burning, and/or dispersants. The secondary strategy is to deter wildlife from entering the oiled areas through techniques such as passive visual methods and active auditory methods (e.g., propane cannons, bird-scare buoys). Preemptive capture and moving of unoiled wildlife is included in the secondary strategy. Only trained and authorized personnel are allowed to conduct active deterrence and capture activities. The tertiary strategy is to capture and rehabilitate oiled wildlife for possible release back to the environment. Rehabilitation and release of wildlife will likely be a key component of oil spill response in the Arctic. Those types of activities are complicated in the Arctic because of logistical issues, the likely remoteness of a spill, the use of many wildlife species for subsistence purposes by local communities, and the size and type of wildlife that might need to be responded to, such as polar bears and walruses. There are also legal issues that would need to be considered for some species, such as adherence to the Marine Mammal Protection Act or the Endangered Species Act.

For many species, federal agencies co-manage with Alaska partners. Rehabilitation and release of wildlife in the event of a spill would need to be accomplished in close association with subsistence hunters and marine mammal co-management organizations, such as the Ice Seal Committee. The Ice

[8] See http://www.oilspillresponse.com/services-landing/well-incident-intervention/subsea-well-response-project.

[9] See http://www.oilandgasuk.co.uk/knowledgecentre/OSPRAG.cfm.

Seal Committee has requested that sick or injured ice seals not be rehabilitated and released because of concerns about possibly introducing diseases into the wild population, but it is not clear what the policy would be in the event of a spill. Historically, Annex G has not utilized traditional knowledge in its response plans; National Oceanic and Atmospheric Administration (NOAA) and other agencies are attempting to include this in the future (presentation by Brad Smith, NOAA, March 2013).

Response methods for birds affected by an oil spill are generally well established (e.g., USFWS, 2003), and deterrent techniques, equipment, and trained personnel are readily available. Alaska Clean Seas owns and maintains a mobile Bird Capture and Stabilization Center and contracts with International Bird Rescue[10] for personnel support. International Bird Rescue also owns a rehabilitation facility in Anchorage.

The U.S. Fish and Wildlife Services (USFWS) is currently updating its 1999 Oil Spill Response Plan for Polar Bears in Alaska (USFWS, 1999, 2013), which will provide more specific information and guidance for polar bear response. USFWS has been attempting to improve coordination between response partners and has been studying how to best clean oiled polar bear fur (USFWS, 2013). Equipment for oiled polar bear response is maintained by Alaska Clean Seas and supplemented by USFWS equipment. However, there is a minimal amount of equipment, with capacity to respond to up to five oiled bears.

NOAA's National Marine Fisheries Service (NMFS) is finalizing revisions to its Marine Mammal Oil Spill Response Guidelines, last revised in 2006.[11] After completion of the revised national guidelines, regional guidelines will be developed, with Alaska as one of the first to be addressed. Deterrence methods for marine mammals have been considered for use during spill drills in Alaska, including air guns or other noisemaking devices. However, no techniques have been tested or are formally approved at this time. This may be of greatest concern for whales and walruses, because there are no approved deterrence methods. There is, however, literature on bowhead whales' avoidance of seismic air guns and other sounds. NOAA faces a number of issues regarding deterrence and rehabilitation of wildlife—a lack of scientific study and developed protocols regarding hazing of marine mammals, working within the existing legal framework, and a lack of consensus with native co-managers. There are presently no facilities in Alaska with capacity to receive and rehabilitate any significant numbers of marine mammals (presentation from Brad Smith, NOAA, March 2013).

Over the past few years Alaska Clean Seas has sponsored an informal North Slope Marine Mammal Response Working Group consisting of the resource agencies (NMFS, USFWS, Alaska Fish and Game), Alaska Zoo, Alaska SeaLife Center, and member company representatives to develop procedures and identify needs for marine mammal response on the north slope of Alaska. Working relationships and procedures are improving, but additional resources are needed to increase capabilities. The Alaska SeaLife Center has developed protocols for the care of oil-affected phocid seals and has reviewed equipment on hand for polar bear response and determined the additional equipment necessary for seal response. They have also designed, constructed, and tested a Mobile Treatment and Rehabilitation Enclosure for seals for Alaska Clean Seas. One positive aspect of response activities in the Arctic is that, unlike other regions, many of the Arctic animals do not need

[10] See http://www.bird-rescue.org/about/our-bird-rescue-centers/alaska-wildlife-response-center.aspx.

[11] See http://www.nmfs.noaa.gov/pr/pdfs/health/eis_appendixl.pdf.

much water for stabilization. Fish totes can be used for most initial activities and a facility with large holding tanks is not necessary.

Planning for marine mammal response in northern Alaska is under way but still in its infancy, with significant logistical challenges. Space, water, power, and personnel support facilities are in short supply, although small portable capabilities are under consideration. Alaska has an extensive stranding network which could assist during a spill event, but trained and authorized live animal capture and handling personnel are in short supply. The Alaska SeaLife Center holds the only NOAA/NMFS permit in Alaska to respond to marine mammal and bird strandings. Support for wildlife response (e.g., boats, personnel, and facilities) is not well defined in contingency plans. Expectations for marine mammal response could be better defined—for example, how to respond to large marine mammals, which are difficult to approach unless they are incapacitated; tactics if an oil spill impacts a haulout; and the consequences of attempting to deter or move an animal versus no response. It does not appear that recent oil spills such as the *Deepwater Horizon* have led to new or different methods of deterrence.

RESPONSE STRATEGIES FOR COASTAL ENVIRONMENTS

A primary goal of a spill response is to keep oil off the coast, in order to minimize impacts to sensitive habitats. The basic oil spill response strategies used in temperate regions are generally applicable to Arctic marine and shoreline environments. Specific differences are likely to relate to the remoteness of Arctic environments and an expectation of challenging working conditions (Potter et al., 2012). As a result, Arctic operations require an understanding of specific response techniques (e.g., mechanical recovery, dispersants, in situ burning), potential limitations associated with each technique, and knowledge of what works with conditions at sea, in ice, or on coastlines during different seasons (Polaris Applied Science, 2009).

General shoreline cleanup approaches include physical removal, in situ techniques, and natural processes (Sergy et al., 2003; Prince et al., 2003a). Additionally, possible interaction between sediments in the shore zone and the oil need to be considered. While there is likely to be little tidal mixing energy, wave action in the surf zone could lead to significant oil-sediment interaction (Owens et al., 2003). Oiled sediments can be treated in different manners, including in situ treatment through dry or wet mixing as well as sediment relocation. The goal in these cases is to increase the exposure of oiled sediments to weathering processes, in order to accelerate natural degradation. In the cases where oil refloats during exposure to wave action or other energetic processes, it can be contained and collected (Polaris Applied Science, 2009). As with most oil spill responses, the Net Environmental Benefit Analysis (NEBA) process (described in the next section) is one approach to assess potential countermeasures, whether they are active (e.g., in situ burning, mechanical cleanup) or passive (monitoring of natural attenuation processes). It also helps to determine whether additional environmental damage could be caused as a result of specific response actions. In remote locations such as the Arctic, where physical access may not be practical, using a biodegradation response may provide for less physical damage than would be incurred by the use of manual removal techniques.

An additional consideration is avoiding the possibility of placing responders in hazardous situations to perform beach cleanup.

The Arctic's remote locations, limited infrastructure availability, and few possibilities for the transport and disposal of oiled waste require response techniques that minimize waste generation. Preferable techniques are in situ and natural attenuation processes, which lead to lower waste generation for shore zone operations. In the case of on-water treatment options, the use of oil/water separators could minimize oily water waste. Onboard incineration of collected oil and oily waste could provide another opportunity to reduce the volumes of contaminated material. New techniques that reduce waste, such as herding agents that allow for in situ burning without the use of fire-resistant boom, may also be of high value for Arctic spills (Buist et al., 2011).

NEBA APPROACH

Regardless of the response method or methods used, there will be some environmental impacts due to an oil spill. Numerous factors must be considered for the selection of oil spill response procedures for use in contingency plans and emergency response operations. These factors include the probability of an oil spill, the possible volume and type of crude oil or refined product that might be spilled, environmental factors influencing the fate and behavior of the hydrocarbons that could be released, the sensitivity of the most valued ecosystem components (VECs) to oil pollution, the potential impacts from the application of oil spill countermeasures (e.g., dispersants, in situ burning), and the time needed for habitat recovery.

The NEBA process became increasingly prominent following the *Exxon Valdez* oil spill. It provides decision makers a strategy for deciding what response options are appropriate at a specific spill location based on the analysis of environmental tradeoffs that may occur from the use of the various oil spill countermeasures available (Baker, 1995). From an ecological point of view, NEBA provides a protocol for weighing the advantages and disadvantages of various spill responses with regard to flora and fauna and their habitats within the specific area of concern, compared with no response (known as natural attenuation). A generic NEBA framework is outlined in the International Petroleum Industry Environmental Conservation Association publication, *Choosing Spill Response Options to Minimize Damage: Net Environmental Benefit Analysis* (IPIECA, 2000). In addition to providing information for the selection of the best cleanup methods, the NEBA process also provides an assessment of long-term effects on an ecosystem, guidance on the intensity level and operational end points for cleanup operations, and estimates of likely recovery rates (Potter et al., 2012; IMO, in press). A decision process such as NEBA can be used to determine which countermeasures might be best to reduce detrimental effects to an already contaminated environment.

Identifying and Protecting Valued Ecosystem Components

The optimal spill response technique is defined as the one that will minimize a spill's adverse impact on the habitat of the region and its biological resources. Case studies have conclusively shown that the application of aggressive cleanup operations may delay the rates of habitat recovery by causing additional damage beyond the oil spill itself (Baker, 1995). For example, in the aftermath of the *Exxon Valdez* incident, excavation and washing of rocks to remove surface and subsurface oil was shown not to offer a net environmental benefit because the procedures altered shore structure and delayed biological recovery (NOAA Hazardous Materials Response Branch, 1990).

In the event of a large oil spill, a single spill response strategy is unlikely to provide optimal protection for all environmental resources as more than one environmental compartment (i.e., water surface, water column, sediments, and shoreline) would likely be impacted. In fact, a response strategy that provides protection for one environmental resource (e.g., chemical dispersion of oil slicks to protect seabirds) may increase risks to another (e.g., toxicity of dispersed oil in the water column). Decision makers select the optimal response strategy based on the protection of priority environmental resources and the countermeasures that offer them the greatest protection.

NEBA incorporates prioritization criteria for the protection of VECs that could be impacted by oiling, cleanup operations, or residual oil. These rare and valuable species of aquatic plants and animals have scientific, social, cultural, economic, historical, archaeological, or aesthetic importance, determined on the basis of cultural ideals or scientific concern (Forbes, 2011). VECs identified for the Chukchi and Beaufort Seas include higher-trophic-level species such as polar bears, whales, and seals; prey species such as capelin and sculpin; and activities such as seal hunts (Lee et al., 2011b). The analysis will also consider seasonal variations of these valued ecosystem components (e.g., breeding grounds, migration routes) and the time frame of the restoration of items which may be impacted.

Using NEBA in the Selection of Oil Spill Countermeasures

There are several response countermeasures available for oil spill response on open water and in ice (physical recovery, dispersant applications, in situ burning, and monitoring natural recovery), as well as on shorelines (e.g., manual oil collection, flushing, cleaning agents, pressure washing, and bioremediation). There are clear differences in operational limits for each oil spill response strategy—advantages for some techniques can include speed, efficiency, or ease of use, while disadvantages can include burn residue or leftover oil due to low encounter rates. The value of a particular method will depend on the situation, including weather conditions, organisms and ecosystems that are impacted, the availability of response support, and the type of oil spilled. A NEBA process needs to include all identified factors in order to select the best response strategy.

Planning a NEBA Strategy for the Arctic

Concerns about accidental releases of oil have risen with the expansion of frontier oil and gas and marine shipping industries within the Arctic Circle (AMAP, 2010). To date, there has not been a large oil spill within Arctic marine waters. Small spills have been primarily associated with fishing activities (AMAP, 2010), although larger spills have occurred in ice-covered subarctic waters such as the Gulf of Finland (*Runner 4*) and the Gulf of St. Lawrence (*Kurdistan, Arrow*). However, with the expected increase in exploration and production operations in deeper offshore waters, and due to

Scenario 6
Well Blowout

An exploratory well in the offshore Chukchi Sea suffers a major blowout late in the drilling season. The shallow well was being drilled in about 45 m of water within a targeted oil zone approximately 100 km from the coast of Alaska. As the blowout continues, the well releases nearly 60,000 bbl/day. The responsible party has abided by all necessary laws and regulations. This represents a low-probability, high-impact event that has potentially large environmental consequences due to the volume of oil released.

Main Considerations
Issues include weather, visibility, ice impacts on response and rescue efforts, potential for needed personnel rescue/recovery and oil spill response, potential for a large volume of oil released, shoreline contamination in remote areas, and oil in ice with cleanup extending over the winter months. Other considerations include the logistics of reaching the well, rescuing personnel, and limiting spill impacts. Oil behavior and its effects are dependent on the season, and response strategies will need to be tailored to meteorological-ocean-ice conditions over a long-term cleanup period.

Standard Response
Responsible party assumes leadership role for cleanup, Unified Command is activated, and efforts are made to make responsible party's response resources immediately available.

Response Needs
Knowledge of long-term behavior of oil in ice, on ice, and under ice; decision tools and processes for rapid and effective response to large-scale, remote, and logistically challenging Spills of National Significance; prepositioning of spill response equipment; resources to cover sociological and environmental damages and for long-term shoreline contamination cleanup; and training and exercises for local response personnel working with the Unified Command personnel, possibly over winter, are needed.

the *Deepwater Horizon* oil spill, there are major concerns over the possibility of a subsurface blowout under or in ice and the industry's and government's ability to respond.

A decision process such as NEBA is likely to be an integral part of future contingency plans, because post-spill decisions are best and most quickly made in light of pre-spill analyses, consultations, and agreements involving all of the appropriate organizations and parties. Use of NEBA in contingency planning offers several advantages—extended time frame for analysis, consideration of spill scenarios covering a wide range of environmental factors (e.g., seasonal changes in species diversity and ice cover), time for identification and collection of scientific data, and stakeholder involvement. NEBA can cover a range of oil spill scenarios from small accidental releases associated with routine operations to the "worst case" in terms of oil volume, sensitive species, and environmental factors. Other important considerations include the logistical constraints that are likely to be encountered during Arctic oil spill response operations, which will influence the efficacy of current

Scenario 7
Structural Failure of an Oil Storage Tank

A land-based diesel oil storage tank suffers structural failure in a winter storm. The oil overruns the berm and flows over the ice and into the offshore environment.

Main Considerations
Response on land, nearshore, and offshore may require different teams, expertise, and coordination (particularly in the offshore area) to account for offshore, shoreline, groundwater, and surface water contamination and contaminant cleanup. Oil in ice and on ice will require cleanup extending over the winter months, with impacts from weather, low visibility, and ice on response efforts. There will be community integration into the response for several reasons: depletion of oil reserves in the winter, without the possibility of restoring reserves until the following spring, and long-term health and environmental risks posed by a tank spill.

Standard Response
Responsible party assumes leadership role for cleanup, Unified Command is activated, and local entities assume leadership within the Unified Command.

Response Needs
Adequate prepositioned response equipment, integrated and simultaneous response on land, nearshore, and offshore, which will need training and exercises for local response personnel working with the Unified Command personnel, provision for fuel and oil replenishment in case of an oil tank leak and/or rupture, wildlife response techniques, and logistical and manpower support for large events are needed.

countermeasure strategies. For an Arctic NEBA, the process needs to include not only regulators, resource managers, health authorities, technical specialists in oil spill response technologies, and the scientific community, but also regional representatives to ensure that local and traditional knowledge is incorporated.

The use of NEBA is helpful for regional spill contingency plans, since some response options have a limited window of opportunity. Contingency planning can identify such areas of potential conflict and attempt to resolve them through consultation among all interested organizations before any spill occurs. While oil spill response protocols have been developed for the Arctic, there is a need to understand the effectiveness and consequences of the technologies such as oil spill treatment agents. NEBA analysis factors in the logistical challenges associated with deploying and supporting oil spill response resources in the Arctic that influence their success. For example, the effectiveness of mechanical oil recovery by skimmers may be highly limited by the presence of ice, while the same ice presence may support in situ burning operations by preventing the spreading of oil.

Conducting a NEBA

Traditional NEBAs involve the following elements (IPIECA, 2000) in the process of collecting information regarding the physical features, ecology, and human resource use in the relevant area. These are to:

1. Review previous spills and experimental results that are relevant to the area and to possible countermeasures;
2. On the basis of prior experience, anticipate likely environmental outcomes if the proposed countermeasure is implemented compared to outcomes if the area is left to recover naturally; and
3. Compare and weigh advantages and disadvantages of various potential responses with those of natural cleanup.

The NEBA process can be used to establish the most important resources at risk in an oil spill, based, for example, on their status as a protected species, ecosystem relevance, or human use. This set of priorities will depend on the preparation of a list of local environments at risk; the results of predictive oil fate, behavior, and trajectory models; and observational and remote sensing data. It takes into account the nature of the spilled oil and changes it may undergo during weathering, which may influence the level of biological effects. Additional information may be available in species vulnerability and environmental sensitivity maps (discussed in Chapter 2), showing the distribution of natural resources and important human use areas.

As discussed briefly in Chapter 3, a "no response" or natural attenuation option is an operational oil spill response strategy. Such an approach is often employed during a spill response because monitoring and evaluation of a situation are basic steps before taking physical cleanup action. NEBA can then be used to provide an estimate of the potential impact on VECs from the spill if no countermeasures were used, based on oil spill trajectory and fate models; oil toxicity for various resource groups, such as coastal marshes, marine mammals, marine birds, and fishes; seasonal and/or spatial distribution and specific characteristics of VECs, such as breeding conditions, habitat, and migration routes; and estimates of impacts to populations and communities of living resources.

For the application of active spill response methodologies, NEBA can determine potential recovery rates for each countermeasure option and can be used to identify possible limits on their practical use. It can also be used to establish which environmental or operational conditions could improve particular response efficiencies. For each type of countermeasure, NEBA ascertains the degree to which environmental impact on each resource could be lessened if the countermeasure was deployed by itself or in combination with others. By considering the reductions in spill impact that could result from each countermeasure, including the time needed to achieve acceptable levels of habitat recovery, the NEBA can be used to select the response options that can offer the best protection to VECs, as well as the greatest overall reduction in harmful environmental consequences.

A decision process such as NEBA can be used to select the intensity of the response countermeasure, as well as the determination of when to conclude the response. Decisions are typically based on potential environmental and socioeconomic impacts of the spill, the potential impact of the cleanup operation, and whether the environment will be able to naturally recover from oil spill impacts. The American Petroleum Institute's (API's) 2012 report, *Spill Response in the Arctic Offshore* (Potter et al., 2012), provides a range of cleanup intensities and conditions where each might be appropriate. They note that the most intense efforts to remove all oil are appropriate in areas where there is an abundance of human or environmental activity, the risk from oil is high, and intense cleanup measures pose little risk. By contrast, monitoring may be sufficient in situations where the oil poses little risk, the potential for natural oil degradation and natural recovery are high, and the risks posed by cleanup efforts are also high. These examples represent end members of cleanup effort. Some situations call for more moderate levels of effort. Another example from their report shows that where areas adjacent to the oiled area are highly sensitive to oil, and where the oiled area is only moderately sensitive to cleaning, the appropriate cleanup efforts may include removing most of the visible oil but allowing traces of oil to remain (Potter et al., 2012).

Data Gaps and Research Needs to Improve the Application of NEBA in the Arctic

Prediction of the fate and effects of oil under various spill scenarios is largely accomplished by the use of spill trajectory models and applicable case histories. The precision and accuracy of these models is based on ocean current data, knowledge of the physical and chemical characteristics of the spilled oil, weather data, and species sensitivity data. In comparison to temperate regions, the database for this information in the Arctic is limited. For example, reliable oil spill trajectory models for oil fate and behavior under sea ice conditions have not been established. Considering its extreme weather conditions, seasonal ice cover, extended periods of darkness, and the possibility of enhanced species sensitivity to oil, the Arctic presents a challenge for the application of NEBA.

Developing standardized protocols for toxicity testing would benefit comparative efforts and support future natural resource damage assessment procedures. For example, risk assessments will require a toxicity database for VECs. There is a need to conduct assessments of native populations and habitat characteristics to establish benchmark conditions at sites slated for future exploration and production and in areas that might be affected by a marine oil spill from shipping. A shift from acute effects to sublethal effects on key Arctic species is needed, including at sensitive life stages. The integration of toxicity data and predictive risk assessment models is essential in future NEBA analysis.

The Arctic environment adds additional levels of uncertainty in the interpretation of toxicity data. Several steps to potentially reduce this uncertainty could be further study of the effects of temperature on the metabolism and biochemical responses of various cold-blooded organisms and studies on the life histories of Arctic organisms to determine seasonal differences in their sensitivity to oil. The results of toxicity studies for temperate species of zooplankton have often been used to infer the response of Arctic species in terms of sensitivity to oil. However, such direct links cannot be drawn. For example, Arctic zooplankton typically have longer life spans, which can lead to increased duration of oil exposure. They also contain higher lipid concentrations, which allow them to survive longer without food and to attain higher concentrations of residual hydrocarbons. There is a need to better understand the impact of oil spills on productivity, the food web, and trophic-level dynamics.

Arctic oil spill case studies have been focused on small response operations, and the Arctic spill response community has limited first-hand experience and thus little information to predict the environmental outcomes of various oil spill countermeasures. Information on the long-term effects of oil and its environmental persistence within the Arctic is limited. Depending on the situation, habitat recovery rates following a spill can be slow. As a result, monitoring habitat recovery with a view toward restoration to pre-spill conditions may not be a valid or achievable end point. Climate change presents an additional confounding factor in the long term.

It is also important to note that there are a number of emerging oil spill countermeasure technologies (e.g., chemical herders, facilitation of oil mineral aggregate formations) that have not been fully validated in the field. Until sufficient information is available on the population dynamics of VECs, the fate and environmental effects of oil, and the efficacy of various response technologies, Arctic NEBAs will be subject to a wider range of response effectiveness and uncertainty in outcomes. The Arctic Oil Spill Response Joint Industry Program has undertaken a project (Environmental Impacts from Arctic Oil Spills and Oil Spill Response Technologies) with a goal to provide information to improve and advance NEBA effectiveness.[12]

While the living resources in the Arctic have high ecological, socioeconomic, and cultural value, the potential loss of ecosystem services, which are the benefits to humankind from resources and processes that are supplied by ecosystems, is another important consideration in the NEBA process. If ecosystem services are lost or damaged within the Arctic, there is a need to determine the value of lost services, their recovery time, and whether they can be replaced by other means.

Socioeconomic and Ecological Considerations in NEBA

There are no response methods that are completely effective or risk free. NEBA considers the advantages and disadvantages of different spill response options, including no response. Prioritizing spill response options usually involves making tradeoffs between environmental resources that have been chosen to receive priority protection. However, socioeconomic factors will play an important part in decisions about spill response, so an optimal technique will minimize a spill's adverse impact on a region's environment, its economy, and the well-being of its people. In all cases, compromises are made during spill response operations. For an area containing both birds and fish, dispersant applications might be the best way to reduce the threat to birds, but could increase the potential exposure level of fish in the area. This could lead to a different set of priorities in the Arctic, where subsistence hunting and fishing are extremely important to the indigenous people.

CHAPTER CONCLUSIONS AND RECOMMENDATIONS

Conclusion: An ongoing effort is needed to determine appropriate wildlife response techniques. In some cases, the "no response" option may be preferable for large marine mammals, such as hauled-out walruses. Wildlife response plans will need to include key indicators of environmental health, and prioritize response strategies.

[12] See http://www.arcticresponsetechnology.org/research-projects/environmental-impacts-from-arctic-oil-spills-and-arctic-oil-spill-response-technologies.

Recommendation: The U.S. Fish and Wildlife Service, NOAA's National Marine Fisheries Service, the Alaska Department of Fish and Game, co-management organizations, and local government and communities are the trustees for wildlife deterrence and rehabilitation. As appropriate, these agencies and groups should work together with industry to explore and improve deterrence and rehabilitation methods for wildlife. Additional research and development for improved methods could benefit from the involvement of universities, non-governmental organizations, and others. Priorities should be set and regularly updated by the trustees for oil spill response based on the type of wildlife threatened, the season, other factors related to a spill, and updated research and methodology.

Conclusion: An Arctic NEBA process requires prioritization of valuable ecosystem components, including seasonal distribution and cultural importance of wildlife, fish, and other resources; information on the transport, fate, and potential effect of the spilled oil; knowledge of operational limits, advantages, and disadvantages of each oil spill response countermeasure for the natural resources at risk; and consideration of logistical constraints and cleanup intensity.

Recommendation: A decision process such as NEBA should be used to select the response options that offer the greatest overall reduction of adverse environmental impacts. In the Arctic, areas of cultural and subsistence importance should be among the priority ecosystem components. In light of concerns regarding detrimental effects on ecosystems, further study should focus on the impact of oil spills on Arctic food webs and dynamics at different trophic levels. The process should involve regulators, resource managers, health authorities, technical specialists, scientific experts, and local experts.

References

ACS (Alaska Clean Seas). 2012. *Alaska Clean Seas Technical Manual, Volume 1: Tactics Description.* ACS, Prudhoe Bay, AK.

Adalsteinsson, D., R. Camassa, S. Harenberg, Z. Lin, R.M. McLaughlin, K. Mertens, J. Reis, W. Schlieper, and B. White. 2013. Subsurface trapping of oil plumes in stratification: Laboratory investigations. Pp. 257-262 in *Monitoring and Modeling the Deepwater Horizon Oil Spill: A Record-Breaking Enterprise*, edited by Y. Liu, A. Macfadyen, Z.-G. Ji, and R. H. Weisberg. American Geophysical Union, Washington, DC.

ADEC (Alaska Department of Environmental Conservation). 2002. Alaska Incident Management System Guide (AIMS) for Oil and Hazardous Substance Response, Revision 1. Available, http://www.akrrt.org/aim/aim_toc.shtml.

ADEC Spill Prevention and Response. 2002a. Alaska Federal/State Preparedness Plan for Response to Oil & Hazardous Substance Discharges/Releases (Unified Plan), Annex G, Wildlife Protection Guidelines.

ADEC Spill Prevention and Response. 2002b. Alaska Federal/State Preparedness Plan for Response to Oil & Hazardous Substance Discharges/Releases (Unified Plan), Annex M, Historic Properties Protection Guidelines for Alaska Federal On-Scene Coordinators.

ADEC Spill Prevention and Response. 2010. Alaska Federal/State Preparedness Plan for Response to Oil & Hazardous Substance Discharges/Releases (Unified Plan), Annex B, Historic Properties Protection Guidelines for Alaska Federal On-Scene Coordinators.

Al-Darbi, M.M., N.O. Saeed, M.R. Islam, and K. Lee. 2005. Biodegradation of natural oils in seawater. *Energy Sources* 27(1-2): 19-34.

Al-Mailem, D.M., N.A. Sorkhoh, H. Al-Awadhi, M. Eliyas, and S.S. Radwan. 2010. Biodegradation of crude oil and pure hydrocarbons by extreme halophilic archaea from hypersaline coasts of the Arabian Gulf. *Extremophiles* 14: 321-328.

Allen, A.A. 1987. Test and evaluation of the helitorch for the ignition of oil slicks. Pp. 243-265 in *Proceedings of the 10th Arctic and Marine Oilspill Program (AMOP) Technical Seminar*. Environment Canada, Ottawa, Ontario.

Allen, A.A. 1990. Contained controlled burning of spilled oil during the *Exxon Valdez* oil spill. Pp. 305-313 in *Proceedings of the 13th Arctic and Marine Oilspill Program (AMOP) Technical Seminar*. Environment Canada, Ottawa, Ontario.

Allen, A.A. 1999. New tools and techniques for controlled in-situ burning. Pp. 613-628 in *Proceedings of the 22nd Arctic and Marine Oilspill Program (AMOP) Technical Seminar*, Volume 2. Environment Canada, Ottawa, Ontario.

Allen, A.A., D.H. Dale, and C. Gregory. 1999. Assessment of potential oil spill recovery capabilities. *International Oil Spill Conference Proceedings* 1999(1): 527-534.

Allen A.A., D. Jaeger, N.J. Mabile, and D. Costanzo. 2011. The use of controlled burning during the Gulf of Mexico Deepwater Horizon MC-252 oil spill response. *International Oil Spill Conference Proceedings* 2011(1): abs194.

Allen, B.M., and R.P. Angliss. 2012. *Alaska Marine Mammal Stock Assessments, 2011*. U.S. Department of Commerce, Seattle, WA, 288 pp.

Alverson, D.L., and N.J. Wilimovsky. 1966. Fishery investigations of the southeastern Chukchi Sea. Pp. 843-860 in *Environment of the Cape Thompson Region, Alaska*, edited by N.J. Wilomovski and J.N. Wolfe. U.S. Atomic Energy Commission, Washington, DC.

AMAP (Arctic Monitoring and Assessment Programme). 2008. Arctic Oil and Gas 2007. AMAP, Oslo, Norway.

AMAP. 2010. Assessment 2007: Oil and Gas Activities in the Arctic–Effects and Potential Effects. Volume 2. AMAP, Oslo, Norway, 277 pp.

Amstrup, S.C. 1995. Movements, distribution, and population dynamics of polar bears in the Beaufort Sea. Ph.D. Dissertation. University of Alaska-Fairbanks, 299 pp.

Amstrup, S.C., G.M. Durner, I. Stirling, N.J. Lunn, and F. Messier. 2000. Movements and distribution of polar bears in the Beaufort Sea. *Canadian Journal of Zoology* 78: 948-966.

Amstrup, S.C., G.M. Durner, T.L. McDonald, D.M. Mulcahy, and G.W. Garner. 2001. Comparing movement patterns of satellite-tagged male and female polar bears. *Canadian Journal of Zoology* 79: 2147-2158.

Anderson, P.J., and J.F. Piatt. 1999. Community reorganization in the Gulf of Alaska following ocean climate regime shift. *Marine Ecology Progress Series* 189:117-123.

Arctic Council. 2009. The Arctic Marine Shipping Assessment 2009 Report. Available, http://www.arctic.noaa.gov/detect/documents/AMSA_2009_Report_2nd_print.pdf.

Arctic Council. 2012. Agreement on Cooperation on Marine Oil Pollution Preparedness and Response in the Arctic. Appendix IV: Operational Guidelines.

Arctic Council. 2013. Recommended Practices for Arctic Oil Spill Prevention. Emergency Prevention, Preparedness and Response.

Aré, F.E. 1988. Thermal abrasion of sea coasts (part 1). *Polar Geography and Geology* 12(1).

Arnborg, L., H.J. Walker, and J. Peippo. 1966. Water discharge in the Colville River, 1962. *Geografiaka Annaler: Series A, Physical Geography* 48(4): 195-210.

Arnborg, L., H.J. Walker, and J. Peippo. 1967. Suspended load in the Colville River, Alaska, 1962. *Geografiaka Annaler: Series A, Physical Geography* 49(2/4): 131-144.

Arrigo, K., D. Perovich, R. Pickart, Z. Brown, G. van Dijken, K. Lowry, M.M. Mills, M.A. Palmer, W.M. Balch, F. Bahr, N.R. Bates, C. Benitez-Nelson, B. Bowler, E. Brownlee, J.K. Ehn, K.E. Frey, R. Garley, S.R. Laney, L. Lubelczyk, J. Mathis, A. Matsuoka, B.G. Mitchell, G.W.K. Moore, E. Ortega-Retuerta, S. Pal, C.M. Polashenski, R.A. Reynolds, B. Schieber, H.M. Sosik, M. Stephens, and J. Swift. 2012. Massive phytoplankton blooms under Arctic sea ice. *Science* 336(6087): 1048.

ARRT (Alaska Regional Response Team). 2008. In-situ burning guidelines for Alaska: Revision 1. Appendix II, Annex F, in *The Alaska Federal/State Preparedness Plan for Response to Oil and Hazardous Substance Discharges/Releases*.

ARRT. 2013. Alaska Regional Response Team Meeting Summary. Available, http://alaskarrt.org/files/ARRTNotes%20 2-21-13.pdf, retrieved 17 August 2013.

ASCE (American Society of Civil Engineers) 1996. Task Committee on Modeling of Oil Spills of the Water Resources Engineering Division. State-of-the-art review of modeling transport and fate of oil spills. *Journal of Hydraulic Engineering* 122(11): 594-609.

ASTM (American Society for Testing and Materials). 2009. ASTM F2152-Standard Guide for In-Situ Burning of Spilled Oil: Fire-Resistant Boom. In *Annual Book of ASTM Standards.*

Atlas, R.M. 1984. *Petroleum Microbiology*. Macmillan, New York, 533 pp.

Atlas, R.M. 1995. Petroleum biodegradation and oil spill bioremediation. *Marine Pollution Bulletin* 31: 178-182.

Atlas, R.M., and R. Bartha. 1972. Biodegradation of petroleum in seawater at low temperatures. *Canadian Journal of Microbiology* 18: 1851-1855.

Atlas, R.M., and R. Bartha. 1992. Hydrocarbon biodegradation and oil-spill bioremediation. *Advances in Microbial Ecology* 12: 287-338.

Atlas, R.M., and T.C. Hazen. 2011. Oil biodegradation and bioremediation: A tale of the two worst spills in U.S. history. *Environmental Science & Technology* 45: 6709-6715.

Babiker, M., K. Kloster, S. Sandven, and R. Hall. 2010. The Utilisation of Satellite Images for the Oil in Ice Experiment in the Barents Sea, May 2009. SINTEF Oil in Ice–JIP, Report No. 29, 44 pp.

Baelum, J., S. Borglin, R. Charkaborty, J.L. Fortney, R. Lamendella, O.U. Mason, and M. Auer, M. Zemla, M. Bill, M.E. Conrad, S.A. Malfatti, S.G. Tringe, H.-Y. Holman, T.C. Hazen, and J.K. Jansson. 2012. Deep-sea bacteria enriched by oil and dispersant from the Deepwater Horizon spill. *Environmental Microbiology* 14(9): 2405-2416.

Baker, J.M. 1995. Net environmental benefit analysis for oil spill response. *International Oil Spill Conference Proceedings*, 1995(1): 611-614.

Bakermans, C., P.W. Bergholz, D.F. Rodrigues, T.A. Vishnivetskaya, H.L. Ayala-del-Rio, and J.M. Tiedje. 2012. Genomic and expression analyses of cold-adapted microorganisms. Pp. 126-155 in *Polar Microbiology: Life in a Deep Freeze*, edited by L. Whyte and R. Miller. ASM Press, Washington, DC.

Banet, A.C. 1994. A Comparison of Crude Oil Chemistry on America's North Slope: Chukchi Sea-Mackenzie Delta. BLM-Alaska Technical Report 17. Bureau of Land Management, Alaska State Office, Anchorage.

Barber, W.E., R.L. Smith, M. Vallarino, and R.M. Meyer. 1997. Demersal fish assemblages of the northeastern Chukchi Sea, Alaska. *Fisheries Bulletin (US)* 95: 195-209.

Barnett, C.J., and J.E. Kontogiannis. 1975. The effect of crude oil fractions on the survival of a tidepool copepod, *Tigriopus californicus*. *Environmental Pollution* 8: 45-54.

Barnett, M., A. Artzer, F. Schmude, D. Diaz, J. Basciani, and G. Harvey. 2012. Generation, evolution and applicable lessons learned from Alaska's historic November 2011 storm over the Bering and Chukchi seas. In *Offshore Arctic Technology Conference*. Houston, TX.

Barnhart, K.R., R.S. Anderson, I. Overeem, C.W. Wobus, G.D. Clow, F.E. Urban, and T.P. Stanton. 2011. Modeling the rate and style of Arctic coastal retreat along the Beaufort Sea, Alaska. *American Geophysical Union Fall Meeting Abstracts* 1: 0803.

Barry, R., R. Moritz, and J. Rogers. 1976. Studies of the climate and fast ice interaction during the decay season along the Alaskan Beaufort Sea coast. Pp. 213-228 in *Proceedings of the 27th Alaska Science Conference. Volume 2*, pp. 213-228.

Barry, R., R. Moritz, and J. Rogers. 1979. The fast ice regimes of the Beaufort and Chukchi Sea coasts. *Cold Regions Science and Technology* 1: 129-152.

Bartha, R. 1986. Biotechnology of petroleum pollutant biodegradation. *Microbial Ecology* 12(1): 155-172.

Baschek, B. 2007. Multi-Sensor Oil Spill Surveillance Program. International Oil & Ice Workshop, Anchorage, AK.

Baumann, U., M. Benz, E. Pletscher, K. Breuker, and R. Zenobi. 1999. Biodegradation of sugar alcohol ethoxylates. *Tenside Surfactants Detergents* 36: 288-293.

Becker, P.R., and C.A. Manen. 1988. Natural Oil Seeps in the Alaskan Marine Environment. Final Report, Outer Continental Shelf Environmental Assessment Program, U.S. Department of Commerce, Technical Information Service, PB88-235965.

Bentzen, R.L., A.N. Powell, and R.S. Suydam. 2009. Strategies for nest site selection by king eiders. *Journal of Wildlife Management* 73(6): 932-938.

Berdugo, V., R.P. Harris, and S.C. O'Hara. 1977. The effect of petroleum hydrocarbons on reproduction of an estuarine planktonic copepod in laboratory cultures. *Marine Pollution Bulletin* 8: 138-143.

Bird, K., J. Charpentier, R. Ronald, D.L. Gautier, D.W. Houseknecht, T.R. Klett, J.K. Pitman, T.E. Moore, C.J. Schenk, M.E. Tennyson, and C.J. Wandrey. 2008. Circum-Arctic resource appraisal; estimates of undiscovered oil and gas north of the Arctic Circle. U.S. Geological Survey Fact Sheet 2008-3049, 4 pp. Available, http://pubs.usgs.gov/fs/2008/3049/, retrieved 29 August 2013.

Bitz, C.M., J.K. Ridley, M. Holland, and H. Cattle. 2012. Global climate models and 20th and 21st century Arctic climate change. Pp. 405-436 in *Arctic Climate Change: The ACSYS Decade and Beyond*, edited by P. Lemke and H.-W. Jacobi. Houten, Springer Netherlands.

Blanchard, A.L., and H.M. Feder. 2003. Adjustment of benthic fauna following sediment disposal at a site with multiple stressors in Port Valdez, Alaska. *Marine Pollution Bulletin* 45: 1590-1599.

Blanchard, A.L., H.M. Feder, and D.G. Shaw. 2003. Variations of benthic fauna underneath an effluent mixing zone at a marine oil terminal in Port Valdez, Alaska. *Marine Pollution Bulletin* 46: 1583-1589.

Blanchard, A.L., H.M. Feder, and M.K. Hoberg. 2010. Temporal variability of benthic communities in an Alaskan glacial fjord, 1971–2007. *Marine Environmental Research* 69: 95-107.

Blanchard, A.L., H.M. Feder, and D.G. Shaw. 2011. Associations between macrofauna and sediment hydrocarbons from treated ballast water effluent at a marine oil terminal in Port Valdez, Alaska. *Environmental Monitoring and Assessment* 178: 461-476.

Blanchard, A.L., C.L. Parris, A.L. Knowlton, and N.R. Wade. 2013. Benthic ecology of the northeastern Chukchi Sea. Part II. Spatial variation of megafaunal community structure, 2009-2010. *Continental Shelf Research* 67: 67-76.

Blenkinsopp, S.A., G.A. Sergy, K.G. Doe, G.D. Wohlgeschaffen, K. Li, and M.F. Fingas. 1997. Evaluation of the toxicity of the weathered crude oil used at the Newfoundland Offshore Burn Experiment (NOBE) and the resultant burn residue. Pp. 677-684 in *Proceedings of the 20th Arctic and Marine Oilspill Program (AMOP) Technical Seminar*. Environment Canada, Ottawa, Ontario.

Blouin, M., and K. Lee. 2007. Oil mineral aggregate formation in ice-infested waters. In *Proceedings of the 2007 International Oil and Ice Workshop*. Minerals Management Service, Herndon, VA.

Bluhm, B.A., and R. Gradinger. 2008. Regional variability in food availability for Arctic marine mammals. *Ecological Applications* 18(2): S77-S96.

Bluhm, B.A., K. Iken, S.M. Hardy, B.I. Sirenko, and B.A. Holladay. 2009. Community structure of epibenthic megafauna in the Chukchi Sea. *Aquatic Biology* 7: 269-293.

Bluhm, B.A., K. Iken, and R.R. Hopcroft. 2010. Observations and exploration of the Arctic's Canada Basin and the Chukchi Sea: The Hidden Ocean and RUSALCA expeditions. *Deep-Sea Research II* 57: 1-4.

BOEM (Bureau of Ocean Energy Management). 2011. Assessment of Undiscovered Technically Recoverable Oil and Gas Resources of the Nation's Outer Continental Shelf, 2011. BOEM Fact Sheet RED-1011-01a.

BOEM. 2013. Alaska Annual Studies Plan FY 2014. U.S. Department of Interior, Bureau of Ocean Energy Management. Anchorage, AK.

Boufadel, M.C., P. Reeser, M.T. Suidan, B.A. Wrenn, J. Cheng, X. Du, T. Huang, and A.D. Venosa. 1999. Optimal nitrate concentration for the biodegradation of n-heptadecane in a variably-saturated sand column. *Environmental Technology* 20(20): 191-199.

Braddock, J.F., M.L. Ruth, P.H. Catterall, J.L. Walworth, and K.A. McCarthy. 1997. Enhancement and inhibition of microbial activity in hydrocarbon-contaminated Arctic soils: Implications for nutrient-amended bioremediation. *Environmental Science & Technology* 31: 2078-2084.

Bradford, J. H., D.F. Dickins, and P.J. Brandvik. 2010. Assessing the potential to detect oil spills in and under snow using airborne ground-penetrating radar. *Geophysics* 75(2): G1-G12.

Bragg, J.R., and S.H. Yang. 1995. Clay-oil flocculation and its effects on the rate of natural cleansing in Prince William Sound following the Exxon Valdez oil spill. Pp. 178-214 in *Exxon Valdez Oil Spill--Fate and Effects in Alaskan Waters*, edited by P.G. Wells, J.N. Butler, and J.S. Hughes. American Society for Testing and Materials, Philadelphia.

Bragg, J.R., R.C. Prince, E.J. Harner, and R.M. Atlas. 1994. Effectiveness of bioremediation for the *Exxon Valdez* oil spill. *Nature* 368: 413-418.

Brakstad, O.G., and K. Bonaunet. 2006. Biodegradation of petroleum hydrocarbons in seawater at low temperatures (0-5 degrees C) and bacterial communities associated with degradation. *Biodegradation* 17: 71-82.

Brakstad, O.G., and L.-G. Faksness. 2000. Biodegradation of water-accommodated fractions and dispersed oil in the seawater column. In *Proceedings of the SPE International Conference on Health Safety and Environment in Oil and Gas Exploration and Production, 26-28 June, 2000, Stavanger, Norway*. Society of Petroleum Engineers, Richardson, TX (CD-ROM).

Brakstad, O.G., I. Nonstad, L.-G. Faksness, and P.J. Brandvik. 2008. Responses of microbial communities in Arctic sea ice after contamination by crude petroleum oil. *Microbial Ecology* 55: 540-552.

Brakstad, O.G., P.S. Daling, L.-G. Faksness, I.K. Almås, S.-H. Vang, and F. Leirvik. 2011. Surface dispersion and depletion of the Macondo MC252 crude oil. In *Proceedings of the 32nd Society of Environmental Toxicology and Chemistry (SETAC) Meeting*, November 13-17, Boston.

Brandvik, P.J., and T. Buvik. 2009. Using Dogs to Detect Oil Hidden in Snow and Ice—Results from Field Training on Svalbard April 2008. SINTEFJIP-rep-no-14-Oildog-snow-ice[1].pdf. Available, http://www.sintef.no/Projectweb/JIP-Oil-In-Ice/Publications/, retrieved 22 August 2011.

Brandvik, P.J., and L.-G. Faksness. 2009. Weathering processes in Arctic oil spills: Meso-scale experiments with different ice conditions. *Cold Regions Science and Technology* 55: 160-166.

Brandvik, P.J., I. Singsaas, and P.S. Daling. 2004. Oil spill R&D in Norwegian Arctic waters with special focus on large-scale oil weathering experiments. In *Proceedings from the Interspill Conference*. Trondheim, Norway.

Brandvik, P.J., L.-G. Faksness, P. Daling, and I. Singsaas. 2005. Fate and behavior of oil spills under Arctic conditions. Earlier results compared with new field experiments on Svalbard. Pp. 584-590 in *AMAP International Symposium on Oil and Gas Activities in the Arctic*, St. Petersburg.

Brandvik, P.J., L.-G. Faksness, D.F. Dickins, and J. Bradford. 2006. Weathering of oil spills under Arctic conditions: Field experiments with different ice conditions followed by in-situ burning. Presented at the 2006 Third Annual NATO/CCMS Oil Spill Response Workshop, October 11-13, Dartmouth, NS, Canada.

Brandvik, P.J., J.L.M. Resby, P.S. Daling, F. Leirvik, and J. Fritt-Rasmussen. 2010. Meso-scale weathering of oil as a function of ice conditions. Oil properties, Dispersibility and In Situ Burnability of Weathered Oil as a Function of Time. SINTEF Materials and Chemistry, Oil in Ice, JIP Report No. 19, 116 pp. Available, http://www.sintef.no/project/JIP_Oil_In_Ice/Dokumenter/publications/JIP-rep-no-19-Common-meso-scale-final.pdf, retrieved 4 October 2013.

Brandvik, P.J., Ø. Johansen, F. Leirvik, U. Farooq, and P.S. Daling. 2013. Droplet breakup in subsurface oil releases–Part 1: Experimental study of droplet breakup and effectiveness of dispersant injection. *Marine Pollution Bulletin* 73(1): 319-326.

Breezee, J., N. Cady, and J.T. Staley. 2004. Subfreezing growth of the sea ice bacterium "*Psychromonas ingrahamii.*" *Microbial Ecology* 47: 300-304.

Brekne, T.M., S. Holmemo, F. Engen, and G. Morten Skeie. 2004. Norwegian Clean Seas Association for Operating Companies (NOFO)—Research and development program for next generation arctic recovery equipment. Presentation at Interspill Conference. Available, http://www.interspill.com/previous-events/2004/pdf/session3/418_BREKNE.pdf.

Bronson, M., E. Thompson, F. McAdams, and J. McHale. 2002. Ice effects on a barge-based oil spill response system in the Alaskan Beaufort Sea. Pp. 1253-1268 in *Proceedings of the 25th Arctic and Marine Oilspill Program (AMOP) Technical Seminar. Environment Canada*, Ottawa, Ontario.

Brothers, L.L.L., P.E. Hart, and C.D.D. Ruppel. 2012. Minimum distribution of subsea ice-bearing permafrost on the US Beaufort Sea continental shelf. *Geophysical Research Letters* 39(15).

Budge, S.M., A.M. Springer, S.J. Iverson, G. Sheffield, and C. Rosa. 2008. Blubber fatty acid composition of bowhead whales, *Balanea mysticetus*: Implication for diet assessment and ecosystem monitoring. *Journal of Experimental Marine Biology and Ecology* 359: 40-46.

Buist, I.A., and D.F. Dickins. 1987. Experimental spills of crude oil in pack ice. Pp. 378-381 in *Proceedings of the 1987 International Oil Spill Conference*. American Petroleum Institute, Washington, DC.

Buist, I.A., S.L. Ross, B.K. Trudel, E. Taylor, T.G. Campbell, P.A. Westphal, M.R. Myers, G.S. Ronzio, A.A. Allen, and A.B. Nordvik. 1994. The Science, Technology and Effects of Controlled Burning of Oil Spills at Sea. MSRC Technical Report Series 94-013, Marine Spill Response Corporation, Washington, DC, 382 pp.

Buist, I.A., B.K. Trudel, J. Morrison, and D.V. Aurand. 1995. Laboratory studies of the physical properties of in-situ burn residues. Pp. 1027-1051 in *Proceedings of the 18th Arctic and Marine Oilspill Program (AMOP) Technical Seminar*, Volume 2. Environment Canada, Ottawa, Ontario.

Buist, I.A., T. Coe, D. Jensen, S. Potter, L. Anderson, K. Bitting, and K. Hansen. 2003a. In-Situ Burn Operations Manual. U.S. Coast Guard Research and Development Center Report CG-D-06-03, Groton, CT.

Buist, I.A., D.F. Dickins, L. Majors, K. Linderman, J. Mullin, and C. Owens. 2003b. Tests to determine the limits to in-situ burning in brash and frazil ice. Pp. 629-648 in *Proceedings of the 26th Arctic and Marine Oilspill Program (AMOP) Technical Seminar*, Volume 2. Environment Canada, Ottawa, Ontario.

Buist, I.A., S. Potter, T. Nedwed, and J. Mullin. 2007. Field research on using oil herding surfactants to thicken oil slicks in pack ice for in-situ burning. Pp. 403-425 in *Proceedings of the 30th Arctic and Marine Oilspill Program (AMOP) Technical Seminar*, Volume 1, Environment Canada, Ottawa, Ontario.

Buist, I.A., R. Belore, D.F. Dickins, A. Guarino, D. Hackenberg, and Z. Wang. 2009. Empirical weathering properties of oil in ice and snow. Pp. 67-107 in *Proceedings of the 32nd Arctic and Marine Oilspill Program (AMOP) Technical Seminar*, Volume 1, Environment Canada, Ottawa, Ontario.

Buist, I.A., S. Potter, and S.E. Sørstrøm. 2010. Barents Sea field test of herder to thicken oil for in-situ burning. Pp. 725-742 in *Proceedings of the 33rd Arctic and Marine Oilspill Program (AMOP) Technical Seminar*, Volume 2. Environment Canada, Ottawa, Ontario.

Buist, I.A., S. Potter, T. Nedwed, and J. Mullin. 2011. Herding surfactants to contract and thicken oil spills in pack ice for *in situ* burning. *Cold Regions Science and Technology* 67: 3-23.

Bunch, J.N., C. Bedard, and T. Cartier. 1983. Abundance and activity of heterotrophic marine bacteria in selected bays at Cape Hatt, N.W.T.: Effects of oil spills, 1981. Canadian Manuscript Report of Fisheries and Aquatic Sciences, No. 1708, Baffin Island Oil Spill (BIOS) project. Canadian Fisheries and Aquatic Sciences, Sainte-Anne-de-Bellevue, PQ, Canada. 96 pp.

Burek, K.A., F.M.D. Gulland, and T.M. O'Hara. 2008. Effects of climate change on Arctic marine mammal health. *Ecological Applications* 18: S126-S134.

Butinar, L., T. Strmole, and N. Gunde-Cimerman. 2011. Relative incidence of ascomycetous yeasts in Arctic coastal environments. *Microbial Ecology* 61: 832-843.

CAFF (Conservation of Arctic Flora and Fauna). 2013. *Arctic Biodiversity Assessment: Report for Policy Makers.* CAFF, Akureyri, Iceland.

Campbell, R.G., E.B. Sherr, C.J. Ashjian, S. Plourde, B.F. Sherr, V. Hill, and D.A. Stockwell. 2009. Mesozooplankton prey preference and grazing impact in the western Arctic Ocean. *Deep-Sea Research, Part II: Topical Studies in Oceanography* 56(17): 1274-1289.

Cater, T.C. 2010. *Tundra Treatment Guidelines: A Manual for Treating Oil and Hazardous Substance Spills to Tundra,* 3rd Edition. Alaska Department of Environmental Conservation, Division of Spill Prevention and Response. Available, http://dec.alaska.gov/spar/perp/r_d/ttman/web/Tundra%20Treatment%20Guidelines%203rd%20Ed.%202010.pdf, retrieved 26 November 2013.

Chang, W., S. Klemm, C. Beaulieu, J. Hawari, L. Whyte, and S. Ghoshal. 2011. Petroleum hydrocarbon biodegradation under seasonal freeze-thaw soil temperature regimes in contaminated soils from a sub-Arctic site. *Environmental Science & Technology* 45: 1061-1066.

CIRCAC (Cook Inlet Regional Citizens Advisory Council). 2012. A Closer Look: Cook Inlet Regional Citizens Advisory Council 2012 Annual Report. Available, http://www.circac.org/images/pdf/2012_annual_report.pdf, retrieved 23 June 2013.

Citta, J.J., L.T. Quakenbush, J.C. George, R.J. Small, M.P. Heide-Jørgensen, H. Brower, B. Adams, and L. Brower. 2012. Winter movements of bowhead whales (*Balaena mysticetus*) in the Bering Sea. *Arctic* 65: 13-34.

Citta, J.J., R.S. Suydam, L.T. Quakenbush, K.J. Frost, and G. O'Corry-Crowe. 2013. Dive behavior of eastern Chukchi Sea beluga whales (*Delphinapterus leucas*), 1998-2008. *Arctic* 66: 389-406.

Clarke, J.T., C.L. Christman, A.A. Brower, and M.C. Ferguson. 2013. Distribution and Relative Abundance of Marine Mammals in the Northeastern Chukchi and Western Beaufort Seas, 2012. Annual Report, OCS Study BOEM 2013-00117. National Marine Mammal Laboratory, Alaska Fisheries Science Center, NMFS, NOAA, Seattle, WA.

Cloutier, D., and B. Doyon. 2008. OMA formation in ice-covered brackish waters: Large-scale experiments. Pp. 71-88 in *Oil Spill Response: A Global Perspective. Proceedings of the NATO CCMS Workshop on Oil Spill Response, Dartmouth, Nova Scotia, 11-13 October 2006*, edited by W.F. Davidson, K. Lee, and A. Cogswell. Springer, Dordrecht, Netherlands.

Cloutier, D., S. Gharbi, and B. Michel. 2005. On the oil-mineral aggregation process: A promising response technology in ice-infested waters. In *2005 International Oil Spill Conference Proceedings*, 2005(1): 527-531.

Comeau, A.M., W.K. Li., J.-E. Tremblay, E.C. Carmack, and C. Lovejoy. 2011. Arctic Ocean microbial community structure before and after the 2007 record ice minimum. *PLoS ONE* 6.

Comfort, G., and W. Purves. 1982. The Behavior of Crude Oil Spilled Under Multi-year Ice. Environmental Protection Service Report EPS 4-EC-82-4. Environment Canada, Ottawa, Ontario.

Cook, L.L., P. Boehm, and R. Barrick. 2011. Degradation patterns of select PAHs provide evidence of photodegradation in surface MC252 crude oil. In *Proceedings of the 32nd Society of Environmental Toxicology and Chemistry (SETAC) Meeting*. Boston.

Coolbaugh, T. 2008. Oil spill detection and remote sensing—an overview with focus on recent events. Pp. 679-691 in *Proceedings of the Arctic and Marine Oilspill Program (AMOP) Technical Seminar,* Volume 2. Environment Canada, Ottawa, Ontario.

Cooper L.W., M. Janout, K.E. Frey, R. Pirtle-Levy, M.L. Guarinello, J.M. Grebmeier, and J.R. Lovvorn. 2011. The relationship between sea ice break-up, water mass variation, chlorophyll biomass, and sedimentation in the northern Bering Sea. *Deep-Sea Research, Part II: Topical Studies in Oceanography* 65-70: 141-162.

Cottrell, M.T., and D.L. Kirchman. 2009. Photoheterotrophic microbes in the Arctic Ocean in summer and winter. *Applied and Environmental Microbiology* 75: 4958-4966.

Couillard, C.M., K. Lee, B. Légaré, and T.L. King. 2005. Effect of dispersant on the composition of the water-accommodated fraction of crude oil and its toxicity to larval marine fish. *Environmental Toxicology of Chemistry* 24: 1496-1504.

Coyle, K.O., B.A. Bluhm, B. Konar, A. Blanchard, and R.C. Highsmith. 2007. Amphipod prey of grey whales in the northern Bering Sea: Changes in biomass and distribution. *Deep-Sea Research, Part II: Topical Studies in Oceanography* 54: 2906-2918.

Curiale, J.A., 1995, Saturated and olefinic terrigenous triterpenoid hydrocarbons in biodegraded tertiary oil of northeast Alaska. *Organic Chemistry* 23: 177-182.

Daling, P.S., P.J. Brandvik, D. Mackay and O. Johansen. 1990. Characterization of crude oils for environmental purposes. Pp. 119-138 in *Proceedings of the 13th Arctic and Marine Oil Spill Program (AMOP) Technical Seminar*. Environment Canada, Ottawa, Ontario.

Daling, P., P.J. Brandvik, F. Leirvik, A. Holumsnes, and C. Rasmussen. 2010. Development and field testing of a flexible system for application of dispersants to oil spills in ice. Pp 787-814 in *Proceedings of the 33rd Arctic and Marine Oilspill Program (AMOP)Technical Seminar.* Environment Canada, Ottawa, Ontario.

Dauvin, J.C. 1982. Impact of *Amoco Cadiz* oil spill on the muddy fine sand *Abra alba* and *Melinna palmata* community from the Bay of Morlaix. *Estuarine, Coastal and Shelf Science* 14: 517-531.

Davidow, W.H., and M.S. Malone. 1992. *The Virtual Corporation: Structuring and Revitalizing the Corporation for the 21st Century*. Edward Burlingame Books/Harper Business Press, New York.

Daykin, M., Ga. Sergy, D. Aurand, G. Shigenaka, Z. Wang, and A. Tang. 1994. Aquatic toxicity resulting from in-situ burning of oil-on-water. Pp. 1165-1193 in *Proceedings of the 17th Arctic and Marine Oilspill Program (AMOP) Technical Seminar*, Environment Canada, Ottawa, Ontario.

Delarue, J., M. Laurinolli, and B. Martin. 2011. Acoustic detection of beluga whales in the northeastern Chukchi Sea, July 2007 to July 2008. *Arctic* 64: 15-24.

Delille, D., and B. Delille. 2000. Field observations on the variability of crude oil impact on indigenous hydrocarbon-degrading bacteria from sub-Antarctic intertidal sediments. *Marine Environmental Research* 49: 403-417.

Delille, D., A. Bassères, and A. Dessommes. 1998. Effectiveness of bioremediation for oil-polluted Antarctic seawater. *Polar Biology* 19: 237-241.Delille, D., E. Pelletier, A. Rodriguez-Blanco, and J-F. Ghiglione. 2009. Effects of nutrient and temperature on degradation of petroleum hydrocarbons in sub-Antarctic coastal seawater. *Polar Biology* 32: 1521-1528.

Delvigne, G.A.L., and C.E. Sweeney. 1988. Natural dispersion of oil. *Oil and Chemical Pollution* 4(4): 281-310.

Deppe, U., H.H. Richnow, W. Michaelis, and G. Antranikian. 2005. Degradation of crude oil by an Arctic microbial consortium. *Extremophiles* 9: 461-470.

DF Dickins Associates Ltd. and Fleet Technology Ltd. 1992. Behaviour of Spilled Oil at Sea (BOSS): Oil-in-Ice Fate and Behaviour. Report prepared for Environment Canada, U.S. Minerals Management Service, and American Petroleum Institute, 200 pp.

DF Dickins Associates Ltd. and OASIS Environmental. 2006. North Slope Nearshore and Offshore Breakup Study. Report prepared for the Alaska Department of Environmental Conservation, Anchorage, 19 pp.

Dickins, D.F., K. Vaudrey, and S. Potter. 2000. A Review of Spill Response, Ice Conditions, Oil Behavior and Monitoring. Oil Spills in Ice Discussion Paper. Prepared by DF Dickins Associates Ltd., Vaudrey & Associates Inc., and SL Ross Environmental Research Ltd. for Alaska Clean Seas, Prudhoe Bay, AK.

DHS (U.S. Department of Homeland Security). 2008. National Incident Management System. Washington, DC. Available, http://www.fema.gov/pdf/emergency/nims/NIMS_core.pdf.

Diaz, E. (ed.) 2008. *Microbial Biodegradation: Genomics and Molecular Biology*. Caister Academic Press, Norfolk, UK. 402 pp.

Dickins, D.F. 1984. *Alaskan Beaufort Sea Ice Atlas*. Prepared for SOHIO Petroleum Company, Dallas, TX.

Dickins, D.F. 2000. Detection and Tracking of Oil Under Ice. Report prepared for the U.S. Department of Interior, Minerals Management Service, Herndon, VA.

Dickins, D.F. 2011. Behavior of oil spills in ice and implications for spill response. In *Proceedings of the Arctic Technology Conference*, Houston, TX, 7-9 February 2011.

Dickins, D.F., and J.H.S. Andersen. 2009. Remote Sensing Technology Review and Screening. Oil in Ice—JIP, Report 22. SINTEF Materials and Chemistry, Marine Environmental Technology. Available, http://www.sintef.no/project/JIP_Oil_In_ Ice/Dokumenter/publications/JIP-rep-no-22- RSScreening-Final.pdf, retrieved 13 November 2013.

Dickins, D.F., and I.A. Buist. 1981. Oil and Gas Under Sea Ice Study: Vols. 1 & 2. Prepared by Dome Petroleum Ltd. for COOSRA, Report CV-1, Calgary, AB, Canada.

Dickins, D.F., and I.A. Buist. 1999. Oil spill countermeasures for ice covered waters. *Journal of Pure and Applied Chemistry* 71(1): L173-L191.

Dickins, D.F., P.J. Brandvik, J. Bradford, L.-G. Faksness, L. Liberty, and R. Daniloff. 2008. Svalbard 2006 experimental oil spill under ice: Remote sensing, oil weathering under Arctic conditions and assessment of oil removal by in-situ burning. Presented at the 2008 International Oil Spill Conference, Savannah, GA.

Dickins, D.F., J.H.S. Andersen, P.J. Brandvik, I. Singsaas, T. Buvik, J. Bradford...and S. Sandven. 2010. Remote Sensing for the Oil Ice Joint Industry Program 2007-2009. Available, http://www.dfdickins.com/pdf/DickinsJIPRemote SensingLR.pdf, retrieved 4 October 2013.

Divoky, G.J. 1984. The pelagic and nearshore birds of the Alaskan Beaufort Sea: Biomass and trophics. Pp. 417-437 in *The Alaska Beaufort Sea: Ecosystems and Environment*, edited by P.R. Barnes, D.M. Schell, and E. Reimnitz. Academic Press, Orlando, FL.

Divoky, G.J. 1987. The Distribution and Abundance of Birds in the Eastern Chukchi Sea in Late Summer and Early Fall. U.S. Minerals Management Service, Alaska OCS Region.

Divoky, G.J. 1998. Factors affecting the growth of a black guillemot colony in northern Alaska. Ph.D. Thesis, University of Alaska Fairbanks, 144 pp.

Divoky, G.J., and B.B. Harter. 2010. Supernormal delay in hatching, embryo cold tolerance and egg-fostering in the black guillemot *Cepphus grylle*. *Marine Ornithology* 38: 7-10.

Djalalova, I., J. Wilczak, S. McKeen, G. Grell, S. Peckham, M. Pagowski, L. DelleMonache, J.T. McQueen, Y. Tang, P. Lee, J. McHenry, W. Gong, V. Bouchet and R. Mathur and R. Mathur. 2010. Ensemble and bias correction techniques for air quality model forecasts of surface O_3 and $PM_{2.5}$ during the TEXAQS-II experiment of 2006. *Atmospheric Environment* 44(4): 455-467.

Douglas, D.C. 2010. Arctic Sea Ice Decline: Projected Changes in Timing and Extent of Sea Ice in the Bering and Chukchi Seas. U.S. Geological Survey, Open-File Report 1176.

Drozdowski, A., S. Nudds, C.G. Hannah, H. Niu, I. Peterson, and W. Perrie. 2011. Review of Oil Spill Trajectory Modelling in the Presence of Ice. Canadian Technical Report of Hydrography and Ocean Sciences 274. Fisheries and Oceans Canada and Bedford Institute of Oceanography, Dartmouth, NS.

Dunton, K.H., T. Weingartner, and E.C. Carmack. 2006. The nearshore western Beaufort Sea ecosystem: Circulation and importance of terrestrial carbon in Arctic coastal food webs. *Progress in Oceanography* 71(2): 362-378.

Dunton, K.H., S.V. Schonberg, and L.W. Cooper. 2012. Food web structure of the Alaskan nearshore shelf and estuarine lagoons of the Beaufort Sea. *Estuaries and Coasts* 35: 416-435.

Durban, J., D. Weller, A. Lang, and W. Perryman. 2013. Estimating Gray Whale Abundance from Shore-Based Counts Using a Multilevel Bayesian Model. Paper SC/65a/BRG02 presented to the Scientific Committee of the International Whaling Commission, 9 pp.

Durner, G.M., S.C. Amstrup, and A.S. Fischbach. 2003. Habitat characteristics of polar bear terrestrial maternal den sites in northern Alaska. *Arctic* 56: 55-62.

Dutta, T.K., and S. Harayama. 2000. Fate of crude oil by the combination of photooxidation and biodegradation. *Environmental Science & Technology* 34: 1500-1505.

Eicken, H.L. 2013. Ocean science: Arctic sea ice needs better forecasts. *Nature* 497(7450): 431-433.

Eicken, H.L., L. Shapiro, A. Gaylord, A. Mahoney, and P. Cotter. 2006. Mapping and Characterization of Recurring Spring Leads and Landfast Ice in the Beaufort and Chukchi Seas. Report AK-03-06. Prepared for Minerals Management Service, Anchorage, AK.

Eid, B.M., and V.J. Cardone. 1992. Beaufort Sea Extreme Waves Study. Environmental Studies Research Funds Report No. 114. Available, http://www.esrfunds.org/pdf/114.pdf, retrieved 27 November 2013.

Eisner, L., N. Hillgruber, E. Martinson, and J. Maselko. 2012. Pelagic fish and zooplankton species assemblages in relation to water mass characteristics in the northern Bering and southeast Chukchi Seas. *Polar Biology* 36: 87-113.

El-Tahan, H., and S. Venkatesh. 1994. Behavior of oil spills in cold and ice-infested waters: Analysis of experimental data on oil spreading. Pp. 337-354 in *Proceedings of the 17th Arctic and Marine Oil Spill Program (AMOP) Technical Seminar*. Environment Canada, Ottawa, Ontario.

EPA (U.S. Environmental Protection Agency). 2011. Screening-Level Hazard Characterization. Crude Oil Category: Sponsored Chemical: Cruise Oil (CASRN 8002-05-9). Available, http://www.epa.gov/chemrtk/hpvis/hazchar/Category_Crude%20Oil_March_2011.pdf, retrieved 11 February 2014.

ExxonMobil Corporation. 2013. Offshore Arctic Oil Spill Prevention, Preparedness and Response. Available, http://www.exxonmobil.com/Corporate/Files/news_pub_2013-arctic-spill-prevent.pdf, retrieved 10 September 2013.

Faksness, L.-G. 2007. Weathering of oil under Arctic conditions. Ph.D. Thesis, University of Bergen, The University in Svalbard.

Faksness, L.-G., and P.J. Brandvik. 2008. Distribution of water soluble components from oil encapsulated in sea ice: Summary of results from three different field seasons. *Cold Region Science and Technology* 54(2): 106-114.

Faksness, L.-G., O.G. Brakstad, M. Reed, F. Leirvik, P.J. Brandvik, and C. Petrich. 2011. Oil in Ice: Transport, Fate and Potential Exposure. Final report submitted to the Coastal Response Research Center. SINTEF Report No. A19275, 126 pp.

Fechhelm, R.G., J.D. Bryan, W.B. Griffiths, B.J. Gallaway, and W.J. Wilson. 1994. The effects of coastal winds on the summer dispersal of young least cisco (*Coregonus sardinella*) from the Colville River to Prudhoe Bay, Alaska: A simulation model. *Canadian Journal of Fisheries and Aquatic Science* 51(4): 890-899.

Fechhelm, R.G., A.M. Baker, B.E. Haley, and M.R. Link. 2009. Year 27 of the long-term monitoring of nearshore Beaufort Sea fishes in the Prudhoe Bay region: 2009 annual report. Report for BP. Exploration (Alaska) Inc., by LGL Alaska Research Associates, Inc., Anchorage, AK, 84 pp.

Feder, H.M., A.S. Naidu, S.C. Jewett, J.M. Hameedi, W.R. Johnson, and T.E. Whitledge. 1994. The northeastern Chukchi Sea: Benthos-environmental interactions. *Marine Ecology: Progress Series (Oldendorf)* 111(1): 171-190.

Federal Interagency Solutions Group. 2010. Oil Budget Calculator—Deepwater Horizon. Report to the National Incident Command. Available, http://www.restorethegulf.gov/sites/default/files/documents/pdf/OilBudgetCalc_Full_HQ-Print_111110.pdf, retrieved 4 October 2013.

Fingas, M.F., and C.E. Brown. 2011. Oil spill remote sensing: A review. Pp. 111-158 in *Oil Spill Science and Technology*, edited by M. Fingas. Gulf Professional Publishing, Burlington, MA.

Fingas, M.F., G. Halley, F. Ackerman, R. Nelson, M.C. Bissonnette, N. Laroche, Z. Wang, P. Lambert, K. Li, P. Jokuty, G. Sergy, E.J. Tennyson, J. Mullin, L. Hannon, R. Turpin, P. Campagna, W. Halley, J. Latour, R. Galarneau, B. Ryan, D.V. Aurand, and R. Hiltabrand. 1995. The Newfoundland Offshore Burn Experiment—NOBE. *International Oil Spill Conference Proceedings* 1995(1): 123-132.

Fingas, M.F., P. Lambert, Z. Wang, K. Li, F. Ackerman, M. Goldthorp, S. Schutz, M. Morganti R. Turpin, R. Nadeau, P. Campagna, and R. Hiltabrand. 2001. Studies of emissions from oil fires. Pp. 767-823 in *Proceedings of the 24th Arctic and Marine Oil Spill Program (AMOP) Technical Seminar*. Environment Canada, Ottawa, Ontario.

Fissel, D.B., J.R. Marko, and M. Martínez de Saavedra Álvarez. 2009. The changing met-ocean and ice conditions in the Beaufort Sea: Implications for offshore oil and gas. In *Proceedings of the 19th International Offshore and Polar Engineering Conference*, Vol. 1, 21-26 June 2009, Osaka, Japan.

Fissel, D.B., H. Melling, D. Billenness, K. Borg, N. Kulan, M. Martinez de Saavedra Álvarez, and A. Slonimer. 2012. The changing wave climate over the Beaufort Sea shelf, 2001-2010. Paper presented at the International Polar Year (IPY 2012) Conference, April, Montreal, QC, Canada.

Foght, J.M., and D.W.S. Westlake. 1982. Effect of the dispersant Corexit 9527 on the microbial degradation of Prudhoe Bay oil. *Canadian Journal of Microbiology* 28: 117-122.

Foght, J.M., P.M. Fedorak, and D.W.S. Westlake. 1983. Effect of the dispersant Corexit 9527 on the microbial degradation of sulfur heterocycles in Prudhoe Bay oil. *Canadian Journal of Microbiology* 29: 623-627.

Forbes, D. L. (ed.). 2011. State of the Arctic Coast 2010—Scientific Review and Outlook. International Arctic Science Committee, Land-Ocean Interactions in the Coastal Zone, Arctic Monitoring and Assessment Programme, International Permafrost Association, Helmholtz-Zentrum, Geesthacht, Germany, 178 pp. Available, http://arcticcoasts.org, retrieved 2 June 2011.

Forsyth, J.V., Y.M. Tsao, and R.D. Blem. 1995. Bioremediation: When is augmentation needed? Pp. 1-14 in *Bioaugmentation for Site Remediation*, edited by R.E. Hinchee, B.C. Alleman, and J. Frederickson. Battelle Press, Columbus, OH.

Francis, J.A., and S.J. Vavrus. 2012. Evidence linking Arctic amplification to extreme weather in mid-latitudes. *Geophysical Research Letters* 39(6).

Francis, O.P., G.G. Panteleev, and D.E. Atkinson. 2011. Ocean wave conditions in the Chukchi Sea from satellite and in-situ observations. *Geophysical Research Letters* 38(24).

Franken, A., and H. Thomsett. 2013. When it takes a network: Creating strategy and agility through wargaming. *California Management Review* 55(3): 107-133.

Fraser, T.P., and M. Wicks III. 1995. Estimation of maximum stable oil droplet sizes at sea resulting from natural dispersion and from use of a dispersant. Pp. 313-316 in *Proceedings of the 18th Arctic and Marine Oilspill Program (AMOP) Technical Seminar*. Environment Canada, Ottawa, Ontario.

French, D., M. Reed, S. Feng, K. Jayko, S. Pavignano, T. Isaji, S. Puckett, A. Keller, F.W. French III, D. Gifford, J. McCue, G. Brown, E. MacDonald, J. Quirk, S. Natzke, R. Bishop, M. Welsh, M. Phillips, B.S. Ingram, and M. Welsh, 1996. The CERCLA Type A Natural Resource Damage Assessment Model for Coastal and Marine Environments (NRDAM/CME), Vol. I. Report to Office of Environmental Affairs, Washington, DC.

Frost, K.J., and L.F. Lowry. 1983. Demersal fishes and invertebrates trawled in the northeastern Chukchi and western Beaufort Seas 1976–1977. U.S. Department of Commerce, NOAA Technical Report NMFS-SSRF-764, 22 pp.

Frost, K.J., L.F. Lowry, and G. Carroll. 1993. Beluga whale and spotted seal use of a coastal lagoon system in the northeastern Chukchi Sea. *Arctic* 46: 8-16.

Gall, A.E., and R.H. Day. 2011. Distribution and Abundance of Seabirds in the Northeastern Chukchi Sea, 2008-2010. Unpublished report to ConocoPhillips, Shell, and Statoil, Anchorage, AK, 74 pp.

Gall, A.E., R.H. Day, and T.J. Weingartner. 2012. Structure and variability of the marine-bird community in the northeastern Chukchi Sea. *Continental Shelf Research* 67: 96-115.

Gallaway, B.J., W.B. Griffiths, P.C. Craig, and J.W. Helmericks. 1983. An assessment of the Colville River delta stock of Arctic cisco–migrants from Canada? *Biological Papers of the University of Alaska* 21: 4-23.

Gallaway, B.J., R.G. Fechhelm, W.B. Griffiths, and J.G. Cole. 1997. Population dynamics of broad whitefish in the Prudhoe Bay region, Alaska. *American Fisheries Society Symposium* 19: 194-207.

Galt, J.A. 1994. Trajectory analysis for oil spills. *Journal of Advanced Marine Technology Conference* 11: 91-126.

Gannon, M.M., and D.A. Holdway. 1999. Metabolic enzyme activities in fish gills as biomarkers of exposure to petroleum hydrocarbons. *Ecotoxicology and Environmental Safety* 44(1): 92-99.

Garcia, M.T., E. Campos, A. Marsal, and I. Ribosa. 2009. Biodegradability and toxicity of sulphonate-based surfactants in aerobic and anaerobic aquatic environments. *Water Research* 43: 295-302.

Gardiner, W.W., J.Q. Word, R.A. Perkins, K.M. McFarlin, B.W. Hester, L.S. Word, and C.M. Ray. 2013. The acute toxicity of chemically and physically dispersed crude oil to key Arctic species under Arctic conditions during the open water season. *Environmental Toxicology and Chemistry* 32(10): 2284-2300.

Garner, G.W., S.T. Knick, and D.C. Douglas. 1990. Seasonal movements of adult female polar bears in the Bering and Chukchi Seas. Pp. 219-226 in *Bears: Their Biology and Management*, Vol. 8, A Selection of Papers from the Eighth International Conference on Bear Research and Management, Victoria, BC, Canada, February 1989. International Association of Bear Research and Management.

Garrett, R.M., I.J. Pickering, C.E. Haith, and R.C. Prince. 1998. Photooxidation of crude oils. *Environmental Science & Technology* 32: 3719-3723.

Garrett, R.M., S.J. Rothenburger, and R.C. Prince. 2003. Biodegradation of fuel oil under laboratory and Arctic marine conditions. *Spill Science and Technology Bulletin* 8: 297-302.

George, J.C. 2009. Growth, morphology, and energetics of bowhead whales (*Balaena mysticetus*). Ph.D. Dissertation, University of Alaska Fairbanks, 186 pp.

Geraci, J.R., and D.J. St. Aubin (eds.). 1990. *Sea Mammals and Oil: Confronting the Risks*. Academic Press, San Diego, CA, 282 pp.

Gerdes, B., and G. Dieckmann. 2005. Biological degradation of crude oil in Arctic sea ice. Presented at the Arctic Operational Platform (ARCOP) Workshop, 19-20 October 2005, St. Petersburg, Russia. Abstract to the report Growth Project GRD-2000-30112 ARCOP Technology and Environment WP 6 Workshop Activities, 16 February 2006.

Ghiglione, J.-F., P.E. Galand, T. Pommier, C. Pedros-Alio, E.W. Maas, K. Bakker, S. Bertilsoni, D.L. Kirchmanj, C. Lovejoyk, P.L. Yagerg, and A.E. Murray. 2012. Pole-to-pole biogeography of surface and deep marine bacterial communities. *Proceedings of the National Academy of Sciences of the United States of America* 109: 17633-17638.

Gilbert, J.R., G.A. Fedoseev, D. Seagars, E. Razlivalov, and A. LaChugin. 1992. Aerial Census of Pacific Walrus, 1990. Marine Mammals Management Technical Report 92-1, U.S. Fish and Wildlife Service, Anchorage, AK.

Givens, G.H., S.L. Edmondson, J.C. George, B. Tudor, R.A. DeLong, and R. Suydam. 2012. Detection probability estimates from the 2011ice-based independent observer survey of the bowhead whales near Barrow, Alaska. Paper SC/64/BRG4 presented to the Scientific Committee of the International Whaling Commission, 25 pp.

Gjøsteen, J.K. 2004. Model for oil spreading in cold waters. *Cold Regions Science and Technology* 38(2-3): 117-125.

Gjøsteen, J.K., S. Loset, and O.T. Gudmestad. 2003. The ability to model oil spills in broken ice. Pp. 717-725 in *Proceedings of the 17th International Conference on Port and Ocean Engineering Under Arctic Conditions*, Volume 2.

Goodman, R. 2008. Oil under ice detection: What is the state of the art? Pp. 7-19 in *Oil Spill Response: A Global Perspective*, edited by W.F. Davidson, K. Lee, and A. Cogswell. Springer Netherlands, Houten.

Goodyear, J., and B. Beach. 2012. Environmental Risks with Proposed Offshore Oil and Gas Development off Alaska's North Slope, Issue Paper: 12-08-A. Natural Resources Defense Council.

Gormezano, L.J., and R.F. Rockwell. 2013. Dietary composition and spatial patterns of polar bear foraging on land in western Hudson Bay. *BMC Ecology* 13: 51.

Gosselin, M., M. Levasseur, P.A. Wheeler, R.A. Horner, and B.C. Booth. 1997. New measurements of phytoplankton and ice algal production in the Arctic Ocean. *Deep-Sea Research, Part II: Topical Studies in Oceanography* 44: 1623-1644.

Grabowski, M., and K.H. Roberts. 1996. Human and organizational error in large-scale systems. *IEEE Transactions on Systems, Man & Cybernetics* 26: 2-16.

Grabowski, M., and K.H. Roberts. 1997. Risk mitigation in large-scale systems: Lessons from high reliability organizations. *California Management Review* 39: 152-162.

Grabowski, M., and K.H. Roberts. 1999. Risk mitigation in virtual organizations. *Organization Science* 10(6): 704-721.

Grabowski, M., and K.H. Roberts. 2011. High-reliability virtual organizations: Co-adaptive technology and organizational structures in tsunami warning systems. *ACM Transactions on Computer-Human Interaction (ToCHI)* 18(4): Article 19.

Grabowski, M., J.H. Harrald, and K.H. Roberts. 1997. Decision support and organizational forms in a high velocity environment: Responses to catastrophic oil spills. In *Advances in Expert Systems for Management*, Vol. 2, edited by M.R. Grabowski and W.A. Wallace. JAI Press, Greenwich, CT.

Grabowski, M., P. Ayyalasomayajula, J.R. Merrick, J.H. Harrald, and K.H. Roberts. 2007. Leading indicators of safety in virtual organizations. *Safety Science* 45(10): 1013-1043.

Gradinger, R., and Q. Zhang. 1997. Vertical distribution of bacteria in Arctic sea ice from the Barents and Laptev Seas. *Polar Biology* 17: 448-454.

Grebmeier, J.M. 2012. Shifting patterns of life in the Pacific Arctic and sub-Arctic seas. *Annual Review of Marine Science* 4: 63-78.

Grebmeier, J.M., H.M. Feder, and C.P. McRoy. 1989. Pelagic-benthic coupling on the shelf of the northern Bering and Chukchi Seas. 2. Benthic community structure. *Marine Ecology Progress Series* 51(3): 253-268.

Grebmeier, J. M., J.E. Overland, S.E. Moore, E.V. Farley, E.C. Carmack, L.W. Cooper, K.E. Frey, J.H. Helle, F.A. McLaughlin, and L. McNutt. 2006. A major ecosystem shift observed in the northern Bering Sea. *Science* 311: 1461-1464.

Grebmeier, J.M., S.E. Moore, J.E. Overland, K.E. Frey, and R. Gradinger. 2010. Biological response to recent Pacific Arctic sea ice retreats. *EOS, Transactions of the American Geophysical Union* 91: 161-168.

Grossman, M., R. Prince, R. Garrett, K. Garrett, R. Bare, K. Lee, G. Sergy, E. Owens, and C. Guénette. 1999. Microbial diversity in oiled and un-oiled shoreline sediments in the Norwegian Arctic. Pp. 775-789 in *Microbial Biosystems: New Frontiers, Proceedings of the 8th International Symposium on Microbial Ecology*, edited by C.R. Bell, M. Brylinsky, and P. Johnson-Green. Atlantic Canada Society for Microbial Ecology, Halifax, NS, Canada.

Guénette, C.C., G.A. Sergy, E.H. Owens, R.C. Prince, and K. Lee. 2003. Experimental design of the Svalbard shoreline field trials. *Spill Science and Technology Bulletin* 8: 245-256.

Gulec, I., and D.A. Holdway. 1999. The toxicity of laboratory burned oil to the amphipod *Allorchestes compressa* and the snail *Polinices conicus. Spill Science and Technology Bulletin* 5: 135-139.

Haller, G., and F.J. Beron-Vera. 2013. Coherent Lagrangian vortices: The black holes of turbulence. *Journal of Fluid Mechanics* 731, 10: R4.

Han, X.M., A.C. Scott, P.M. Fedorak, M. Bataineh, and J.W. Martin. 2008. Influence of molecular structure on the biodegradability of naphthenic acids. *Environmental Science & Technology* 42: 1290-1295.

Hansen, K., and M. Lewandowski. 2011. Using exercises to advance approaches to response for oil in ice environments. Arctic Technology Conference, OTC Paper 22122-MS, February 2011, Houston.

Harper, J.R., R.F. Henry, and G.G. Stewart. 1988. Maximum storm surge elevations in the Tuktoyaktuk region of the Canadian Beaufort Sea. *Arctic* 48-52.

Harrald, J.R., R. Cohn, and W.A. Wallace. 1992. We were always reorganizing…some crisis management implications of the Exxon Valdez oil spill. *Industrial Crisis Quarterly* 6: 197-217.

Harter, B.B. 2007. Black guillemots as indicators of change in the near-shore Arctic marine ecosystem. M.S. Thesis, University of Manitoba, Winnipeg, Manitoba, Canada, 129 pp.

Hartwell, A.D. 1973. Classification and relief characteristics of northern Alaska's coastal zone. *Arctic* 26(3): 244-252.

Harwood, L.A., and M.C.S. Kingsley. 2013. Trends in the offshore distribution and relative abundance of Beaufort Sea belugas, 1982-85 vs 2007-09. *Arctic* 66: 247-256.

Harwood, L.A., S. Innes, P. Norton, and M.C.S. Kingsley. 1996. Distribution and abundance of beluga whales in the Mackenzie Estuary, southeast Beaufort Sea and west Amundsen Gulf during late July 1992. *Canadian Journal of Fisheries and Aquatic Sciences* 53: 2262-2273.

Harwood, L.A., F. McLaughlin, R.M. Allen, J. Illasiak, Jr., and J. Alikamik. 2005. First-ever marine mammal and bird observations in the deep Canada Basin and Beaufort/Chukchi Seas: Expeditions during 2002. *Polar Biology* 28: 250-253.

Hazen, T.C., E.A. Dubinsky, T.Z. DeSantis, G.L. Anderson, Y.M. Piceno, N. Singh, J.K. Jansson, A. Probst, S.E. Borglin, J.L. Fortney, W.T. Stringfellow, M. Bill, M.E. Conrad, L.M. Tom, K.L. Chavarria, T.R. Alusi, R. Lamendella, D.C. Joyner, C. Spier, J. Baelum, M. Auer, M.L. Zemla, R. Chakraborty, E.L. Sonnenthal, P. D'haeseleer, H.-Y.N. Holman, S. Osman, Z. Lu, J.D. Van Nostrand, Y. Deng, J. Zhou, and O.U. Mason. 2010. Deepsea oil plume enriches indigenous oil-degrading bacteria. *Science* 330: 204-208.

Heiss-Blanquet, S., Y. Benoit, C. Maréchaux, and F. Monot. 2005. Assessing the role of alkane hydroxylase genotypes in environmental samples by competitive PCR. *Journal of Applied Microbiology* 99: 1392-1403.

Hibler, W.D., III. 1979. A dynamic thermodynamic sea ice model. *Journal of Physical Oceanography* 9: 815-846.

Hills, S., and J.R. Gilbert. 1994. Detecting Pacific walrus population trends with aerial survey—a review. *Transactions of the North American Wildlife and Natural Resource Conference* 59: 201-210.

Hirvi, J.P., R. Hakala, and J. Rytkonen. 1987. Study of Crude Oil Behavior in Sub-Arctic Brackish Water. Report by the Finnish National Board of Waters and the Environment and the Technical Research Centre of Finland, Helsinki.

Hodgins, D.O., S.S. Salvador, S.E. Tinis, and D. Nazarenko. 1996. Radarsat SAR for oil spill response. *Spill Science and Technology Bulletin* 3(4): 241-246.

Holland-Bartels, L., and B. Pierce (eds.). 2011. An Evaluation of the Science Needs to Inform Decisions on Outer Continental Shelf Energy Development in the Chukchi and Beaufort Seas, Alaska. U.S. Geological Survey Circular 1370, 278 pp. Available, http://pubs.usgs.gov/circ/1370/pdf/circ1370.pdf.

Hopcroft, R.R., and R.H. Day. 2013. Introduction to the special issue on the ecology of the northeastern Chukchi Sea. *Continental Shelf Research* 67: 1-4.

Hopcroft, R.R., K.N. Kosobokova, and A.I. Pinchuk. 2010. Zooplankton community patterns in the Chukchi Sea during summer 2004. *Deep-Sea Research, Part II: Topical Studies in Oceanography* 57: 27–39.

Hopkins, M.A. 1996. On the mesoscale interaction of lead ice and floes. *Journal of Geophysical Research* 101(C8): 18315-18326.

Hopkins, M., and S.F. Daly. 2003. Recent advances in discrete element modeling of river ice. CGU HS Committee on River Ice Processes and the Environment, 12th Workshop on the Hydraulics of Ice Covered Rivers. Edmonton, AB, Canada. Available, http://cripe.civil.ualberta.ca/Downloads/12th_Workshop/Hopkins-Daly-2003.pdf.

Huang, J.C. 1983. A review of the state-of-the-art of oil spill fate/behavior models. *Oil Spill Conference Proceedings* 1983(1): 313-322.

Humphrey, B., P.D. Boehm, M.C. Hamilton, and R.J. Norstrom. 1987. The fate of chemically dispersed and untreated crude oil in Arctic benthic biota. *Arctic* 40(1): 149-161.

Hunt, G.L., Jr., A.L. Blanchard, P. Boveng, P. Dalpadado, K.F. Drinkwater, L. Eisner, R.R. Hopcroft, K.M. Kovacs, B.L. Norcross, P. Renaud, M. Reigstad, M. Renner, H. Rune Skjoldal, A. Whitehouse, and R.A. Woodgate. 2013. The Barents and Chukchi Seas: Comparison of two Arctic shelf ecosystems. *Journal of Marine Systems* 109-110: 43-68.

Hunter, C.M., H. Caswell, M.C. Runge, E.V. Regehr, S.C. Amstrup, and I. Stirling. 2007. Polar Bears in the Southern Beaufort Sea II: Demography and Population Growth in Relation to Sea Ice Conditions. *U.S. Geological Survey, Alaska Science Center, Administrative Report.* 46 pp.

Hutchings, J.K., and I. Rigor. 2012. Role of ice dynamics in anomalous ice conditions in the Beaufort Sea during 2006 and 2007. *Journal of Geophysical Research* 117(C00E04): 14.

Iken, K., B. Bluhm, and K. Dunton. 2010. Benthic food-web structure under differing water mass properties in the southern Chukchi Sea. *Deep-Sea Research, Part II: Topical Studies in Oceanography* 57: 71-85.

IMO (International Maritime Organization). 2013 in press. *Guidelines for the Use of Chemical Dispersants for Combatting Oil Pollution at Sea.* London.

IPCC (Intergovernmental Panel on Climate Change). 2007. Summary for policymakers. In *Climate Change 2007: The Physical Science Basis.* Contribution of Working Group I to the Fourth Assessment Report of the Intergovernmental Panel on Climate Change, edited by S. Solomon, D. Qin, M. Manning, Z. Chen, M. Marquis, K.B. Averyt, M.Tignor, and H.L. Miller. Cambridge University Press, Cambridge, UK, and New York.

IPCC. 2013. Working Group I Contribution to the IPCC Fifth Assessment Report (AR5), Final Draft Climate Change 2013: The Physical Science Basis. Underlying Scientific-Technical Assessment. WG-I: 12th/ Doc. 2b, Add.1. (22IX.2013). Stockholm, Sweden. Available, http://www.climatechange2013.org/images/uploads/WGIAR5_WGI-12Doc2b_FinalDraft_All.pdf, retrieved 5 November 2013.

IPCC. 2014. Climate Change 2014: Impacts, Adaptation, and Vulnerabilities, IPCC Working Group II Contribution to the Fifth Assessment Report (AR5). Available, http://ipcc-wg2.gov/AR5/, retrieved 14 October 2014.

IPIECA (International Petroleum Industry Environmental Conservation Association). 2000. *Choosing Spill Response Options to Minimize Damage Net Environmental Benefit Analysis.* London.

ITOPF (International Tanker Owners Pollution Federation Limited). 2013/2014. *Handbook. In the Event of a Spill of Oil or Hazardous and Noxious Substance (Chemical).* Impact PR & Design Limited, London.

Ivanov, B., A. Bezgreshnov, N. Kubyshkin, and A. Kursheva. 2005. Spreading of oil products in sea ice and their influence on the radiation properties of the snow-ice cover. Pp. 853-862 in *Proceedings of the 18th International Port and Oceans Engineering Under Arctic Conditions.* Clarkson University, Potsdam.

Jay, C.V., A.S. Fischbach, and A.A. Kochnev. 2012. Walrus areas of use in the Chukchi Sea during sparse sea ice cover. *Marine Ecology Progress Series* 468(14): 1-13.

Jeffries, M.O., and J. Richter-Menge (eds.). 2013. The Arctic [in "State of the Climate in 2012"]. *Bulletin of the American Meteorological Society* 94(8): S111-S132.

Jensen, L.K., and J. Carroll. 2010. Experimental studies of reproduction and feeding for two Arctic-dwelling *Calanus* species exposed to crude oil. *Aquatic Biology* 10: 261-271.

Johansen, Ø., and K. Skognes. 1995. Oil Drift in Ice Model. Oceanor Report OCN 95026 to Offshore Operators Committee. Nord, Stavanger, Norway. 23 pp. + appendies.

Johansen, Ø., P.J. Brandvik, and U. Farooq. 2013. Droplet breakup in subsea oil releases. Part 2: Predictions of droplet size distributions with and without injection of chemical dispersants. *Marine Pollution Board* 73(1): 327-335.

Johnson, L. 1997. Living with uncertainty. Pp. 340-345 in *Fish Ecology in Arctic North America: Proceedings of the Fish Ecology in Arctic North America Symposium,* edited by J.B. Reynolds. American Fisheries Society.

Johnson, M., H. Eicken, M.L. Druckenmiller, and R. Glenn (eds.). 2013. Experts Workshops to Comparatively Evaluate Coastal Currents and Ice Movement in the Northeastern Chukchi Sea; Barrow and Wainwright, Alaska. University of Alaska Fairbanks, 37 pp.

Johnson, S.R. 1993. An important early-autumn staging area for Pacific flyway brant: Kasegaluk Lagoon, Chukchi Sea, AK. *Journal of Field Ornithology* 64: 539-548.

Johnson, S.R., D.A. Wiggins, and P.F. Wainwright. 1993. Late-summer abundance and distribution of marine birds in Kasegaluk Lagoon, Chukchi Sea, AK. *Arctic* 46: 212-227.

Johnston, M. 2004. Properties of Second Year and Multi-year Ice at Freeze-up. Canadian NRC--Canadian Hydraulics Centre Technical Report CHC-TR-024.

Jones, B.M., K.M. Hinkel, C.D. Arp, and W.R. Eisner. 2008. Modern erosion rates and loss of coastal features and sites, Beaufort Sea coastline, Alaska. *Arctic* 61: 361-372.

Jones, B.M., C.D. Arp, R.A. Beck, G. Grosse, J.M. Webster, and F.E. Urban. 2009a. Erosional history of Cape Halkett and contemporary monitoring of bluff retreat, Beaufort Sea coast, Alaska. *Polar Geography* 32: 129-142.

Jones, B.M., C.D. Arp, M.T. Jorgenson, K.M. Hinkel, J.A. Schmutz, and P.L. Flint. 2009b. Increase in the rate and uniformity of coastline erosion in Arctic Alaska. *Geophysical Research Letters* 36(3): L03503.

Jones, R.K., 1997. A simplified pseudo-component oil evaporation model. Pp. 43-61 in *Proceedings of the Twentieth Arctic and Marine Oilspill Program (AMOP) Technical Seminar*, Environment Canada, Emergencies Science Division, Ottawa, Ontario.

Jorgenson, M.T., V. Romanovsky, J. Harden, Y. Shur, J. O'Donnell, E.A.G. Schuur, M. Kanevskiy, and S. Marchenko. 2010. Resilience and vulnerability of permafrost to climate change. *Canadian Journal of Forest Research* 40: 1219-1236.

Junge, K., J. Imhoff, J. Staley, and J. Deming. 2002. Phylogenetic diversity of numerically important bacteria in Arctic sea ice. *Microbial Ecology* 43: 315-328.

Junge, K., H. Eicken, and J. W. Deming. 2003. Motility of *Colwellia psychrerythraea* strain 34H at subzero temperatures. *Applied and Environmental Microbiology* 69: 4282-4284.

Junge, K., H. Eicken, and J. W. Deming. 2004. Bacterial activity at -2 and -20°C in Arctic wintertime sea ice. *Applied and Environmental Microbiology* 70: 550-557.

Junge, K., H. Eicken, B.D. Swanson, and J.W. Deming. 2006. Bacterial incorporation of leucine into protein down to -20°C with evidence for potential activity in sub-eutectic saline formations. *Cryobiology* 52: 417-429.

Kay, J. E., M.M. Holland, C.M. Bitz, E. Blanchard-Wrigglesworth, A. Gettelman, A. Conley, and D. Bailey. 2012. The influence of local feedbacks and northward heat transport on the equilibrium Arctic climate response to increased greenhouse gas forcing. *Journal of Climate* 25(16): 5433-5450.

Khelifa, A., L.O. Ajijolaiya, P. MacPherson, K. Lee, P.S. Hill, S. Gharbi, and M. Blouin. 2005. Validation of OMA formation in cold brackish and sea waters. Pp. 527-38 in *Proceedings of the 28th Arctic and Marine Oilspill Program (AMOP) Technical Seminar on Environmental Contamination and Response, June 7-9, Calgary, Alberta.* Environment Canada, Ottawa, Ontario.

Kimes, N.E., A.V. Callaghan, D.F. Aktas, W.L. Smith, J. Sunner, B.T. Golding, M. Drozdowska, T.C. Hazen, J.M. Suflita, and P.L. Morris. 2013. Metagenomic analysis and metabolite profiling of deep-sea sediments from the Gulf of Mexico following the Deepwater Horizon oil spill. *Frontiers in Microbiology* 4(50): 1-17.

Kirchman, D.L., V. Hill, M.T. Cottrell, R. Grandinger, R.R. Malmstrom, and A. Parker. 2009. Standing stocks, production, and respiration of phytoplankton and heterotrophic bacteria in the western Arctic Ocean. *Deep Sea Research, Part II: Topical Studies in Oceanography* 56(17): 1237-1248.

Kjeilen-Eilertsen, G., J.M. Jersak, and S. Westerlund. 2011. Developing treatment products for increased microbial degradation of petroleum oil spills across open-water surfaces. Presented at the 2011 Arctic Technology Conference, Houston, TX, 9 pp.

Kobayashi, N., J.C. Vidrine, R.B. Nairn and S.M. Soloman. 1999. Erosion of frozen cliffs due to storm surge on Beaufort Sea Coast. *Journal of Coastal Research* 15: 332-334.

Konrad, J., and T. Shroder. 2011. *Fire on the Horizon: The Untold Story of the Gulf Oil Disaster*. Harper Collins, New York.

Kovacs, A. 1976. Grounded Ice in the Fast Ice Zone Along the Beaufort Coast of Alaska. U.S. Army Corps of Engineers Cold Regions Research and Engineering Laboratory Library Report 76-32, Hanover, NH.

Krishfield, R., J. Toole, A. Proshutinsky, and M.-L. Timmermans. 2008. Automated ice-tethered profilers for seawater observations under pack ice in all seasons. *Journal of Atmospheric and Oceanic Technology* 25(11): 2091-2105.

Krishfield, R., A. Proshutinsky, K. Tateyama, W.J. Williams, E.C. Carmack, F.A. McLaughlin, and M.-L. Timmermans. 2014. Deterioration of perennial sea ice in the Beaufort Gyre from 2003 to 2012 and its impact on the oceanic freshwater cycle. *Journal of Geophysical Research, Oceans*, 119(2): 1271-1305.

Kujawinski, E.B., M.C. Kido Soule, D.L. Valentine, A.K. Boysen, K. Longnecker, and M.C. Redmond. 2011. Fate of dispersants associated with the *Deepwater Horizon* oil spill. *Environmental Science & Technology* 45: 1298-1306.

Kvenvolden, K.A., and C.K. Cooper. 2003. Natural seepage of crude oil into the marine environment. *Geophysical Marine Letters* 23: 140-146.

Kwok, R., and G.F. Cunningham. 2010. Contribution of melt in the Beaufort Sea to the decline in Arctic multiyear sea ice coverage: 1993-2009. *Geophysical Research Letters* 37(20): L20501.

Labay, K.A., and P.J. Haeussler. 2008. Combined high-resolution LIDAR topography and multibeam bathymetry for northern Resurrection Bay, Seward Alaska. *U.S. Geological Survey Data Series* 374: 6.

LaBelle, J.C., J.L.Wise, R.P. Voelker, R.H. Schulze, and G.M. Wohl. 1983. *Alaska Marine Ice Atlas*. Arctic Environmental Information and Data Center, University of Alaska Anchorage.

Laidre, K.L., I. Stirling, L.F. Lowry, O. Wiig, M.P. Heide-Jorgensen, and S. Ferguson. 2008. Quantifying the sensitivity of Arctic marine mammals to climate-induced habitat change. *Ecological Applications* 18(2 Supplement): S97-S125.

Lampela, K. 2007. Baltic approach to oil spill recovery in ice: Case studies and recent development in Baltic Sea states. In *Proceedings International Oil & Ice Workshop 2007*. Minerals Management Service, Herndon, VA.

Landes, K. 1973. Mother Nature as an oil polluter. *American Association of Petroleum Geologists (AAPG) Bulletin* 57: 637-641.

Lantuit, H., P.P. Overduin, N. Couture, S. Wetterich, F. Are, D. Atkinson, A. Graves-Gaylord, M. Grigoriev, H.-W. Hubberten, J. Jordan, T. Jorgenson, R.S. Ødegård, S. Ogorodov, W.H. Pollard, V. Rachold, S. Sedenko, S. Solomon, F. Steenhuisen, I. Streletskaya, and A. Vasiliev. 2011. The Arctic Coastal Dynamics Database: A new classification scheme and statistics on Arctic permafrost coastlines. *Estuaries and Coasts* 35: 383-400.

Larned, W., R. Stehn, and R. Platte. 2012. Waterfowl breeding population survey, Arctic Coastal Plain, Alaska, 2011. Unpublished report, U.S. Fish and Wildlife Service, Division of Migratory Bird Management, Soldotna, AK, 53 pp.

Leahy, J.G., and R.R. Colwell. 1990. Microbial degradation of hydrocarbons in the environment. *Microbiology and Molecular Biology Reviews* 54: 305-315.

Lee, K., and S. de Mora. 1999. In-situ bioremediation strategies for oiled shoreline environments. *Environmental Technology* 20: 783-794.

Lee, K., C.S. Wong, W.J. Cretney, F.A. Whitney, T.R. Parsons, C.M. Lalli, and J. Wu. 1985. Microbial response to crude oil and Corexit 9527: SEAFLUXES enclosure study. *Microbial Ecology* 11: 337-351.

Lee, K., G.H. Tremblay, and E.M. Levy. 1993. Bioremediation: Application of slow-release fertilizers on low-energy shorelines. *International Oil Spill Proceedings* 1993(1): 449-454.

Lee, K., G.H. Tremblay, J. Gauthier, S.E. Cobanli, and M. Griffin. 1997. Bioaugmentation and biostimulation: A paradox between laboratory and field results. *International Oil Spill Conference Proceedings* 1997(1): 697-705.

Lee, K., P. Stoffyn-Egli, G. Wohlgeschaffen, J. Gauthier, S. St.-Pierre, G.H. Tremblay, S.E. Cobanli, R. Prince, R.E. Bare, R.M. Garrett, M.J. Grossman, G. Sergy, C.C. Guénette, and B.T. Johnson. 1998a. In Situ Treatment of Oiled Sediment Shorelines: Environmental Impact and Validation of Oil Fine Interactions. TTOSS Committee Management Report, 35 pp.

Lee, K., P. Stoffyn-Egli, P. Wood, and T. Lunel. 1998b. Formation and structure of oil-mineral fines aggregates in coastal environments. Pp. 911-921 in *Proceedings of the 21st Arctic and Marine Oilspill Program (AMOP) Technical Seminar*, Environment Canada, Ottawa, Ontario.

Lee, K., P. Stoffyn-Egli, and E.H. Owens. 2001. Natural dispersion of oil in a freshwater ecosystem: Desaguadero Pipeline Spill, Bolivia. *International Oil Spill Conference* Proceedings, 2001(2): 1445-1448.

Lee, K., P. Stoffyn-Egli, G.H. Tremblay, E.H. Owens, G.A. Sergy, C.C. Guénette, and R.C. Prince. 2003a. Oil-mineral aggregate formation on oiled beaches: Natural attenuation and sediment relocation. *Spill Science and Technology Bulletin* 8: 285-296.

Lee, K., G. Wohlgeschaffen, G.H. Tremblay, B.T. Johnson, G.A. Sergy, R.C. Prince, C.C. Guénette, and E.H. Owens. 2003b. Toxicity evaluation with the Microtox test to assess the impact of in-situ oiled shoreline treatment options: Natural attenuation and sediment relocation. *Spill Science and Technology Bulletin* 8: 273-284.

Lee, K., A.D. Venosa, and F.X. Merlin. 2005. Marine oil spill bioremediation: Field studies for development of operational spill response strategies. In *Proceedings of the International Council for the Exploration of the Sea (ICES) CM.* Theme Session on Oil Spills in Marine Ecosystems: Impacts and Remediation. The Council, Copenhagen, Denmark.

Lee, S.H., D.M. Schell, T.L. McDonald, and W.J. Richardson. 2005. Regional and seasonal feeding by bowhead whales *Balaena mysticetus* as indicated by stable isotope ratios. *Marine Ecology Progress Series* 285: 271-287.

Lee, K., Z. Li, H. Niu, P. Kepkay, Y. Zheng, M. Boufadel, and Z. Chen. 2009a. Enhancement of Oil-Mineral-Aggregate Formation to Mitigate Oil Spills in Offshore Oil and Gas Activities. Final Report submitted to Minerals Management Service, 99 pp.

Lee, K., Z. Li, B. Robinson, P.E. Kepkay, X. Ma, S. Cobanli, T. King, M. Blouin, and B. Doyon. 2009b. In-situ remediation of oil spills in ice-infested waters: Enhanced dispersion and biodegradation of petroleum hydrocarbons. In *Proceedings of the 10th International In Situ and On-Site Bioremediation Symposium, May 5-8, 2009.* Battelle Press, Baltimore, MD.

Lee, K., Z. Li, B. Robinson, P.E. Kepkay, M. Blouin, and B. Doyon. 2011a. Field trials of in-situ oil spill countermeasures in ice-infested waters. Presented at the 2011 International Oil Spill Conference, Portland, Oregon. American Petroleum Institute, Washington, DC, 16 pp.

Lee, K., Z. Li, B. Robinson, P.E. Kepkay, M. Blouin, and B. Doyon. 2011b. Oil spill countermeasures in the Arctic. In Pp. 93-108 in *Proceedings of the International Conference on Oil Spill Risk Management: Preparedness, Response and Contingency Planning in the Shipping and Offshore Industries*, edited by N. Bellefontaine and O. Linden. WMU Publications.

Lee, K., Z. Li, B. Robinson, P.E. Kepkay, M. Blouin and B. Doyon. 2012. Oil spill countermeasures in the Arctic. In *Proceedings of the WMU-IMO Conference on Oil Spill Risk Management*, 7-9 March 2011, Neil Bellefontaine and Olof Linden (Eds.). World Maritime University Publications, Malmo Borshus, Sweden. 93-108 pp.

Lee, K., T. Nedwed, R.C. Prince, and D. Palandro. 2013. Lab tests on the biodegradation of chemically dispersed oil should consider the rapid dilution that occurs at sea. *Marine Pollution Bulletin* 73: 314-318.

Leifer, I., W.J. Lehr, D. Simecek-Beatty, E. Bradley, R. Clark, P. Dennison, Y. Hu, S. Matheson, C.E. Jones, B. Holt, M. Reif, D.A. Roberts, J. Svejkovsky, G. Swayze, and J.I. Wozencraft. 2012. State of the art satellite and airborne marine oil spill remote sensing: Application to the BP Deepwater Horizon oil spill. *Remote Sensing of Environment* 124: 185-209.

Lesack, L.F., and P. Marsh. 2007. Lengthening plus shortening of river-to-lake connection times in the Mackenzie River Delta respectively via two global change mechanisms along the Arctic coast. *Geophysical Research Letters* 34(23): L23404.

Lewis, E.L. 1976. Oil in Sea Ice. Pacific Marine Science Report 76-12. Institute of Ocean Science, Patricia Bay, Victoria, BC, Canada.

LGL Alaska Research Associates, Inc., JASCO Applied Sciences, Inc., and Greeneridge Sciences, Inc. 2013. Joint Monitoring Program in the Chukchi and Beaufort Seas, 2013. GL Alaska Draft Report P1272-2 for Shell Offshore, Inc., ION Geophysical, Inc., and other industry contributors, 320 pp. + appendies.

Li, W.K.W., F.A. McLaughlin, C. Lovejoy, and E.C. Carmack. 2009. Smallest algae thrive as the Arctic Ocean freshens. *Science* 326(5952): 539.

Li, Z.K., P. Kepkay, K. Lee, T. King, M.C. Boufadel, A.D. Venosa. 2007. Effects of chemical dispersants and mineral fines on crude oil dispersion in a wave tank under breaking waves. *Marine Pollution Bulletin* 54: 983-993.

Liebezeit, J., and S. Zack. 2010. Avian Habitat and Nesting Use in the Northeast Region of the National Petroleum Reserve—Alaska, Ikpikpuk River Site—2010 Report. Unpublished report of the Wildlife Conservation Society, Portland, OR, 37 pp.

Lillis, P.G., M.D. Lewan, A. Warden, S.M. Monk, and J.D. King. 1999. Identification and characterization of oil types and their source rocks, Chapter OA (Oil Analyses), The Oil and Gas Resource Potential of the 1002 Area, Arctic National Wildlife Refuge, Alaska, by ANWR Assessment Team, U.S. Geological Survey Open-File Report 98-34, Pp. OA-1 to OA-101.

Lindstrom, J.E., and J.F. Braddock. 2002. Biodegradation of petroleum hydrocarbons at low temperature in the presence of the dispersant Corexit 9500. *Marine Pollution Bulletin* 44: 739-747.

Liu, D. 1983. Fate of oil dispersants in aquatic environment. *Science of the Total Environment* 32: 93-98.

Loseto, L.L., G.A. Stern, T.L. Connelly, D. Deibel, B. Gemmill, A. Prokopowicz, L. Fortier, and S.H. Ferguson. 2009. Summer diet of beluga whales inferred by fatty acid analysis of the eastern Beaufort Sea food web. *Journal of Experimental and Marine Biology and Ecology* 374: 12-18.

Lovley, D.R., J.D. Coates, D.A. Saffarini, and D.J. Lonergan. 1997. Dissimilatory iron reduction. pp. 187-215 in *Transition Metals in Microbial Metabolism*, edited by G. Winkelmann and C.J. Carrano. Harwood Academic Publishers, Geneva.

Lowry, L.F., K.J. Frost, and J.J. Burns. 1980. Feeding of bearded seals in the Bering and Chukchi Seas and trophic interaction with Pacific walruses. *Arctic* 33: 330-342.

Lowry, L.F., K.J. Frost, R. Davis, R.S. Suydam, and D.P. DeMaster. 1998. Movements of satellite tagged spotted seals (*Phoca largha*) in the Bering and Chukchi Seas, 1991-1994. *Polar Biology* 19: 221-230.

Lowry, L.F., V.N. Burkanov, K.J. Frost, M.A. Simpkins, R. Davis, D.P. DeMaster, R. Suydam, and A. Springer. 2000. Habitat use and habitat selection by spotted seals (*Phoca largha*) in the Bering Sea. *Canadian Journal of Zoology* 78: 1959-1971.

Lowry, L.F., G. Sheffield, and J.C. George. 2004. Bowhead whale feeding in the Alaskan Beaufort Sea, based on stomach contents analyses. *Journal of Cetacean Research and Management* 6: 215-223.

Lunel, T., K. Lee, R. Swannell, P. Wood, J. Rusin, N. Bailey, C. Halliwell, L. Davies, M. Sommerville, A. Dobie, D. Mitchell, and M. McDonagh. 1996. Shoreline cleanup during the *Sea Empress* incident: The role of surf washing (clay-oil flocculation), dispersants and bioremediation. Pp. 1521-1540 in *Proceedings of the 19th Arctic and Marine Oilspill Program (AMOP) Technical Seminar*. Environment Canada, Ottawa, Ontario.

Lunel, T., L. Davies, S. Shimwell, V. Byfield, S. Boxall and C. Gurney. 1997. Review of aerial/satellite remote sensing carried out at the *Sea Empress* incident. In *Proceedings of the Third International Airborne Remote Sensing Conference*, Copenhagen, Denmark.

Lysne, L.A., E.J. Mallek, and C.P. Dau. 2004. Near Shore Surveys of Alaska's Arctic Coasts, 1999-2003. Unpublished report, U.S. Fish and Wildlife Service, Migratory Bird Management, Waterfowl Branch, Fairbanks, AK. 60 pp.

Mabile, N.J. 2010. Fire Boom Performance Evaluation: Controlled Burning During the Deepwater Horizon Oil Spill. Operational Period, April 28th to July 18th 2010. BP America.

Macdonald, R.W., E. Sakshaug, and R. Stein. 2004. The Arctic Ocean: Modern status and recent climate change. Pp. 6-21 in *The Organic Carbon Cycle in the Arctic Ocean*, edited by R. Stein and R.W. Macdonald. Springer-Verlag, Berlin.

Macnaughton, S.J., R. Swannell, F. Daniel, and L. Bristow. 2003. Biodegradation of dispersed Forties crude and Alaskan North Slope oils in microcosms under simulated marine conditions. *Spill Science and Technology Bulletin* 8: 179-186.

Mageau, C., F.R. Engelhardt, E.S. Gilfillan, and P.D. Boehm. 1987. Effects of short-term exposure to dispersed oil in Arctic invertebrates. *Arctic* 40(1): 162-171.

Mahoney, A. 2012. Sea Ice Conditions in the Chukchi and Beaufort Seas. Prepared by the University of Alaska Fairbanks for the Pew Charitable Trusts U.S. Arctic Program.

Mahoney, A., H. Eicken, L.H. Shapiro, R. Gens, J. Heinrichs, F.J. Meyer and A. Graves. 2012. A Mapping and Characterization of Recurring Spring Leads and Landfast Ice in the Beaufort and Chukchi Seas. OCS Study BOEM 2012-067 prepared for USDOI, BOEM, Alaska OCS Region, 199 pp.

Mahoney, A., H. Eicken, and L. Shapiro. 2007. How fast is landfast ice? A study of the attachment and detachment of nearshore ice at Barrow, AK. *Cold Regions Science and Technology* 47: 233-255.

Maki, H., T. Sasaki, and S. Harayama. 2001. Photo-oxidation of biodegraded crude oil and toxicity of the photo-oxidized products. *Chemosphere* 44: 1145-1151.

Manley, T.O., and K. Hunkins. 1985. Mesoscale eddies of the Arctic Ocean. *Journal of Geophysical Research* 90: 4911-4930.

Manson, G.K., and S.M Solomon. 2007. Past and future forcing of Beaufort Sea coastal change. *Atmosphere-Ocean* 45(2): 107-122.

Marcario, J.C. 2013. Papp: Growing risks in Arctic demand Coast Guard's attention. *Seapower Magazine*. Available, http://seapowermagazine.org/stories/20130521-arctic.html, retrieved 29 August 2013.

Margesin, R., S. Gander, G. Zacke, A.M. Gounot, and F. Schinner. 2003. Hydrocarbon degradation and enzyme activities of cold-adapted bacteria and yeasts. *Extremophiles* 7: 451-458.

Marianoa, A.J., V.H. Kourafalou, A. Srinivasan, H. Kang, G.R. Halliwell, E.H. Ryand, and M. Roffer. 2011. On the modeling of the 2010 Gulf of Mexico oil spill. *Dynamics of Atmospheres and Oceans* 52: 322-340.

Marine Exchange of Alaska. 2009-2012 Report of Recorded Transits. Available, http://www.mxak.org/vtrack/login.html, retrieved 14 November 2013.

Markus, T., J.C. Stroeve, and J. Miller. 2009. Recent changes in Arctic sea ice melt onset, freeze-up, and melt season length. *Journal of Geophysical Research* 114: C12024.

Mars, J.C., and D.W. Houseknecht. 2007. Quantitative remote sensing study indicates doubling of coastal erosion rate in past 50 yr along a segment of the Arctic coast of Alaska. *Geology* 35: 583-586.

Maslanik, J.A., C. Fowler, J. Stroeve, S. Drobot, J. Zwally, D. Yi, and W. Emery. 2007. A younger, thinner Arctic, ice cover: Increased potential for rapid, extensive sea-ice loss. *Geophysical Research Letters* 34(24).

Masterson, D.M. 2009. State of the art of ice bearing capacity and ice construction. *Cold Regions Science and Technology* 58(3) 99-112.

Matsuno, K., A. Yamaguchi, T. Hirawake, and I. Imai. 2011. Year-to-year changes of the mesozooplankton community in the Chukchi Sea during summers of 1991, 1992 and 2007, 2008. *Polar Biology* 34: 1349-1360.

McConnell, I. 2013. Russian relationships: Existing cooperation and future opportunities in the Coast Guard. *Coast Guard Journal of Safety & Security at Sea, Proceedings of the Marine Safety and Security Council* 70(2): 72-76.

McDonald, T.L., W.J. Richardson, C.R. Greene, Jr., S.B. Blackwell, C.S. Nations, R.M. Nielson, and B. Streever. 2012. Detecting changes in the distribution of calling bowhead whales exposed to fluctuating anthropogenic sounds. *Journal of Cetacean Research Management* 12: 91-106.

McFarlin, K.M., R.C. Prince, R. Perkins, and M.B. Leigh. 2014. Biodegradation of dispersed oil in Arctic seawater at −1°C. *PLoS ONE* 9(1).

McGrattan, K.B., W.D. Walton, A.D. Putorti, W.H. Twilley, J. McElroy, and D.D. Evans. 1995. Smoke plume trajectory from in-situ burning of crude oil in Alaska—field experiments. Pp. 901-913 in *Proceedings of the 18th Arctic and Marine Oilspill Program (AMOP) Technical Seminar*, Volume 2. Environment Canada, Ottawa, Ontario.

McLeod, W.R., and D.L. McLeod. 1972. Measures to combat offshore Arctic oil spills. Offshore Technology Conference Paper 1523.

McMinn, T.J. 1972. Crude Oil Behavior on Arctic Winter Ice: Final Report. U.S. Coast Guard, Office of Research & Development, 56 pp.

Mecklenburg, C.W., T.A. Mecklenburg, and L.K. Thorsteinson. 2002. Fishes of Alaska. *American Fisheries Society*, Bethesda, MD, 1037 pp.

Melling, H., and D.A. Riedel. 2004. Draft and Movement of Pack Ice in the Beaufort Sea: A Time-Series Presentation April 1990–August 1999. Canadian Technical Report of Hydrography and Ocean Sciences 238, Institute of Ocean Sciences, Sidney, BC, Canada.

Melling, H., R. Francois, P.G. Myers, W. Perrie, A. Rochon, and R.L.Taylor. 2012. The Arctic Ocean—A Canadian Perspective. *Climatic Change* 115: 89-114.

Meredith, W., S.J. Kelland, and D.M. Jones. 2000. Influence of biodegradation on crude oil acidity and carboxylic acid composition. *Organic Geochemistry* 31: 1059-1073.

Michaud, L., A. Lo Giudice, M. Saitta, M. De Domenico, and V. Bruni. 2004. The biodegradation efficiency on diesel oil by two psychotrophic Antarctic marine bacteria during a two-month-long experiment. *Marine Pollution Bulletin* 49: 405-409.

Mikkola, H. and Kääplyä J. 2013. Arctic Economic Potential: The Need for a Comprehensive and Risk-Aware Understanding of Arctic Dynamics. FIIA Briefing Paper 127. The Finnish Institute of International Affairs.

Miller, M.C., V. Alexander, and R.J. Barsdate. 1978. The effects of oil spills on phytoplankton in an Arctic lake and ponds. *Arctic* 31: 192-218.

Moitoret, C.S., and R.S. Suydam. 1997. Kasegaluk Lagoon Breeding Bird Surveys, 1996 Progress Report. NAES-PR-97-01, U.S. Fish and Wildlife Service, Fairbanks, AK.

Moline, M.A., N.J. Karnovsky, Z. Brown, G.J. Divoky, T.R. Frazer, C.A. Jacoby, J.J. Torres, and W.R. Fraser. 2008. High latitude changes in ice dynamics and their impact on polar marine ecosystems. *The Year in Ecology and Conservation Biology 2008*, edited by R.S. Ostfeld and W.H. Schlesinger. *Annals of the New York Academy of Sciences* 1134 (Special Issue): 267-319.

Moore, S.E. 2000. Variability of cetacean distribution and habitat selection in the Alaskan Arctic, autumn 1982-91. *Arctic* 54: 448-460.

Moore, S.E., and H.P. Huntington. 2008. Arctic marine mammals and climate change: Impacts and resilience. *Ecological Applications* 18(2 Supplement): S157-S165.

Moore, S.E., J.C. George, K.O. Coyle, and T. Weingartner. 1995. Bowhead whales along the Chukotka Coast in autumn. *Arctic* 48(2): 155-160.

Moore, S.E., D.P. DeMaster, and P.K. Dayton. 2000. Cetacean habitat selection in the Alaskan Arctic during summer and autumn. *Arctic* 53: 432-447.

Morison, J., R. Kwok, C. Peralta-Ferriz, M. Alkire, I. Rigor, R. Andersen, and M. Steele. 2012. Changing Arctic Ocean freshwater pathways. *Nature* 481(7379): 66-70.

Moulton, L.L., B. Seavy, and J. Pausanna. 2010. History of an under-ice subsistence fishery for Arctic cicso and least cisco in the Colliver River, AK. *Arctic* 63: 381-390.

Mrinalini P. Nikrad, M.T. Cottrell, and D.L. Kirchman. 2012. Abundance of Single-Cell Activity of Heterotrophic Bacterial Groups in the Western Arctic Ocean in Summer and Winter. *Applied Environmental Microbiology* 78(7): 2402-2409.

Mullin, J. 2012. The oil and gas industry's commitment to responsible arctic operations: An innovative arctic oil spill response technology—Joint Industry Program. OTC Paper 23800, presented at the Offshore Technology Conference, Houston, TX.

Mullin, J., H.V. Jensen, and W. Cox. 2003. MORICE: New technology for effective oil recovery in ice. *International Oil Spill Conference Proceedings 2003(1): 821-825.*

Muschenheim, D.K., and K. Lee. 2003. Removal of oil from the sea surface through particulate interactions: Review and prospectus. *Spill Science and Technology Bulletin* 8: 9-18.

Mykytczuk, N.C.S., S.J. Foote, C.R. Omelon, G. Southam, C.W. Greer, and L.G. Whyte. 2013. Bacterial growth at −15°C: Molecular insights from the permafrost bacterium *Planococcus halocryophilus* Or1. *International Society for Microbial Ecology* 7: 1211-1226.

National Commission on the BP Deepwater Horizon Oil Spill and Offshore Drilling. 2011. Deepwater: The Gulf Oil Disaster and the Future of Offshore Drilling--Report to the President (BP Oil Spill Commission Report). National Commission on the BP Deepwater Horizon Oil Spill and Offshore Drilling.

Nedwed, T. 2010. New dispersant gel treats marine oil spills more effectively with less product. In *Proceedings of the SPE International Conference on Health, Safety and Environment in Oil and Gas Exploration and Production, 12-14 April 2010, Rio de Janeiro, Brazil.* Society of Petroleum Engineers. Available, http://www.onepetro.org/mslib/servlet/onepetropreview?id=SPE-127085-MS, retrieved 28 October 2013.

Nedwed, T., W. Spring, R. Belore, and D. Blanchet. 2007. Basin-scale testing of ASD icebreaker enhanced chemical dispersion of oil spills. Pp. 151-160 in *Proceedings of the 30th Arctic and Marine Oilspill Program (AMOP) Technical Seminar*, Volume 1. Environment Canada, Ottawa, Ontario.

Nedwed, T., G.P. Canevari, J.R. Clark, and R. Belore. 2008. New dispersant delivered as a gel. In *Proceedings of the 2008 International Oil Spill Conference.* American Petroleum Institute, Washington, DC.

Nedwed, T., G.P. Canevari, R. Belore, J. Clark, T. Coolbaugh, and A. Tidwell. 2011. New dispersant gel effective on cold, viscous oils. Poster: *International Oil Spill Conference Proceedings* 2011(1): abs109.

Nelson, W.G., and A.A. Allen, 1982. The physical interaction and cleanup of crude oil with slush and solid first year ice. Pp. 37-59 in *Proceedings of the 5th Arctic and Marine Oilspill Program (AMOP) Technical Seminar*. Environment Canada, Ottawa, Ontario.

Nichols, W.J., and A.D. Venosa. 2008. Summary of the literature on the use of commercial bioremediation agents for cleanup of oil-contaminated environments. *International Oil Spill Conference Proceedings*, 2008(1): 1275-1280.

Nielson, R.M, T.J. Evans, and M.B. Stahl. 2013. Investigating the potential use of aerial line transect surveys for estimating polar bear abundance in sea ice habitats: A case study for the Chukchi Sea. *Marine Mammal Science* 29: 389-406.

Nilsson, A.E., M. Sommerkorn, C. Wilkinson, M. Robards, and T. Vlasova. 2013. The Arctic Resilience Report: Background Aims and Scope, in Arctic Council 2013 Arctic Resilience Interim Report. Stockholm Environment Institute and Stockholm Resilience Center, Stockholm. Available, http://www.arctic-council.org/arr, retrieved 20 May 2013.

Ni'matuzahroh, M., M. Gilewicz, M. Guiliano, and J.C. Bertrand. 1999. In-vitro study of interaction between photooxidation and biodegradation of 2-methylphenanthrene by *Sphingomonas* sp. 2MPII.Chemosphere 38(11): 2501-2507.

NOAA (National Oceanic and Atmospheric Administration). 2011. Automated Data Inquiry for Oil Spills 2. ADIOS2.

NOAA (National Oceanic and Atmospheric Administration) Gulf Spill Restoration. 2012. Natural Resource Damage Assessment, April 2012 Status Update for the *Deepwater Horizon* Oil Spill. Available, http://www.gulfspillrestoration.noaa.gov/wp-content/uploads/FINAL_NRDA_StatusUpdate_April2012.pdf, retrieved 13 November 2013.

NOAA Hazardous Materials Response Branch. 1990. Excavation and Rock Washing Treatment Technology: Net Environmental Benefit Analysis. NOAA, Seattle, WA.

NOAA National Marine Fisheries Service. 2012. Alaska Fisheries Science Center Quarterly Report: October November December 2012.

NOAA Office of Coast Survey, Marine Chart Division. 2013. Arctic Nautical Charting Plan: A Plan to Support Sustainable Transportation in Alaska and the Arctic.

NORCOR Engineering & Research Ltd. 1975. The Interaction of Crude Oil with Arctic Sea Ice. Beaufort Sea Project Technical Report No. 27, Prepared for Department of the Environment, Victoria, BC, Canada.

Norcross, B.L., B.A. Holladay, and C.W. Mecklenburg. 2013a. Recent and Historical Distribution and Ecology of Demersal Fishes in the Chukchi Sea Planning Area. Final Report to the Coastal Marine Institute, Task Order M07AC12462, OCS Study BOEM 2012-073, Fairbanks, AK, 200 pp.

Norcross, B.L., S.W. Raborn, B.A. Holladay, B.J. Gallaway, S.T. Crawford, J.T. Priest, L.E. Edenfield, and R. Meyer. 2013b. Northeastern Chukchi Sea demersal fishes and associated environmental characteristics, 2009–2010. *Continental Shelf Research* 67: 77-95.

Norcross, B.L., L.E. Edenfield, B.P. Gray, B.A. Holladay and K.L. Walker. 2014. Central Beaufort Sea Marine Fish Monitoring, Final Report. U.S. Department of the Interior, Bureau of Ocean Energy Management, Alaska Outer Continental Shelf Region, Anchorage.

NRC (National Research Council). 1989. *Using Oil Spill Dispersants on the Sea*. National Academy Press, Washington, DC.

NRC. 1994. *Committee on Nautical Charts and Information. Charting a Course into the Digital Era*. National Academy Press, Washington, DC.

NRC. 1996. *Understanding Risk: Informing Decisions in a Democratic Society*. National Academy Press, Washington, DC.

NRC. 2003. *Oil in the Sea III: Inputs, Fates and Effects*. National Academies Press, Washington, DC.

NRC. 2005. *Oil Spill Dispersants: Efficacy and Effects*. National Academies Press, Washington, DC.

NRC. 2006. *Toward an Integrated Arctic Observing Network*. National Academies Press, Washington, DC.

NRC. 2011. *National Security Implications for Climate Change for U.S. Naval Forces*. National Academies Press, Washington, DC.

NRC. 2013. *An Ecosystem Services Approach to Assessing the Impacts of the Deepwater Horizon Oil Spill in the Gulf of Mexico*. National Academies Press, Washington, DC.

NRC. 2014. *Linkages Between Arctic Warming and Mid-Latitude Weather Patterns: Summary of a Workshop*. National Academies Press, Washington, DC.

NREL (National Renewable Energy Laboratory). 2009. Biodiesel Handling and Use Guide 4th Edition. NREL/TP-540-43672. Available, http://www.nrel.gov/vehiclesandfuels/pdfs/43672.pdf, retrieved 6 September 2013.

NSTC (National Science and Technology Council). 2013. Arctic Research Plan: FY2013-2017.

NWAB (Northwest Arctic Borough) Subsistence Mapping Project. 2011. Conference Report: November 2-4, 2011, Kotzebue. Available, http://www.nwabor.org/forms/subsistencemapconfreport.pdf, retrieved 30 August 2013.

Oceana. 2008. Arctic: As goes the Arctic, so goes the planet. *Oceana*.

Odokuma, L.O., and G.C. Okpokwasili. 1992. Role of composition in degradability of oil spill dispersants. *Waste Management* 12: 39-43.

OGP (International Association of Oil & Gas Producers). 2011. Oil Spill Response: Global Industry Response Group Recommendations. Report No. 465.

OGP and IPIECA (International Petroleum Industry Environmental Conservation Association). 2012. Current Status and Future Industry Needs for Aerial Dispersant Application: Final Report. Oil Spill Response Joint Industry Programme.

O'Hara, T.M., M.M. Krahn, D. Boyd, P.R. Becker, and L.M. Philo. 1999. Organochlorine contaminant levels in Eskimo harvested bowhead whales of Arctic Alaska. *Journal of Wildlife Disease* 35: 741-752.

Ohmsett. 2013. "Ice Month": Testing in Arctic conditions. *The Ohmsett Gazette*. Spring/Summer 2013.

Ohmsett. 2014. Cold-water comparison of dispersant effectiveness. *The Ohmsett Gazette*. Spring 2014.

Okkonen, S., and T. Weingartner. 2003. Nearshore circulation on the Alaskan Beaufort shelf. P. 24 in *MMS-Beaufort Sea Physical Oceanography Workshop Proceedings*. U.S. Department of the Interior, Minerals Management Service. Available, http://www.ims.uaf.edu/beaufort/FinalPhysicalOceanographyBFProceedings.pdf.

Oppel, S., D.L. Dickson, and A.N. Powell. 2009. International importance of the eastern Chukchi Sea as a staging area for migrating king eiders. *Polar Biology* 32: 775-783.

OSCA (Oil Spill Commission Action). 2013. Assessing Progress Three Years Later. Available, http://oscaction.org/wp-content/uploads/FINAL_OSCA-No2-booklet-Apr-2013_web.pdf, retrieved 29 August 2013.

Osterkamp. T.E. 2007. Characteristics of the recent warming of permafrost in Alaska. *Journal of Geophysical Research* 112F2: F02S02.

Osterkamp, T.E., and W.D. Harrison. 1982. Temperature measurements in subsea permafrost off the coast of Alaska. Pp. 238-248 in *4th Canadian Permafrost Conference*. Available, http://136.159.35.81/cpc/CPC4-238.pdf, retrieved 25 October 2013.

Overland, J.E., and M. Wang. 2013. When will the summer Arctic be nearly sea ice free? *Geophysical Research Letters* 40(10): 2097-2101.

Overland, J.E., J.A. Francis, E. Hanna, and M. Wang. 2012. The recent shift in early summer Arctic atmospheric circulation. *Geophysical Research Letters* 39(19): L19804.

Overland, J.E., J. Key, B.-M. Kim, S.-J. Kim, Y. Liu, J. Walsh, M. Wang, U. Bhatt, and R. Thoman. 2013. [The Arctic] Air temperature, atmospheric circulation, and clouds [in "State of the Climate in 2012"]. *Bulletin of the American Meteorological Society* 94(8): S111-S113.

Ovsienko, S., S. Zatsepa, and A. Ivchenko. 1999. Study and modelling of behavior and spreading of oil in cold water and in ice conditions. Pp. 848-857 in *Proceeding of the 15th International Conference on Port and Ocean Engineering Under Artic Conditions*. Helsinki University of Technology, Ship Laboratory, Espoo, Finland.

Owens, E.H. 1999. The interaction of fine particles with stranded oil. *Pure and Applied Chemistry* 71: 83-93.

Owens, E.H., and K. Lee. 2003. Interaction of oil and mineral fines on shorelines: Review and assessment. *Marine Pollution Bulletin* 47: 397-405.

Owens, E.H., L.B. Solsberg, M.R. West, and M. McGrath. 1998. *Field Guide for Oil Spill Response in Arctic Waters*. Prepared for the Emergency Prevention, Preparedness and Response Working Group. Available, http://arctic-council.org/eppr/wp-content/uploads/2010/04/fldguide.pdf.

Owens, E.H., G.A. Sergy, C.C. Guénette, R.C. Prince, and K. Lee. 2003. The reduction of stranded oil by in-situ shoreline treatment options. *Spill Science and Technology Bulletin* 8: 257-272.

Payne, J.R., G.D. McNabb, Jr., and J.R. Clayton, Jr. 1991. Oil weathering behavior in Arctic environments. *Polar Research* 10(2): 631-662. Available, http://www.polarresearch.net/index.php/polar/article/viewFile/6774/7607, retrieved 26 August 2013.

Pelletier, E., D. Delille, and B. Delille. 2004. Crude oil bioremediation in sub-Antarctic intertidal sediments: Chemistry and toxicity of oiled residue. *Marine Environmental Research* 57: 311-327.

Perovich, D.K., J.A. Richter Menge, K.F. Jones, and B. Light. 2008. Sunlight, water, and ice: Extreme Arctic sea ice melt during the summer of 2007. *Geophysical Research Letters* 35(11): L11501.

Perovich, D.K., K.F. Jones, B. Light, H. Eicken, T. Markus, J. Stroeve, and R. Lindsay. 2011. Solar partitioning in a changing Arctic sea-ice cover. *Annals of Glaciology* 52(57): 192-196.

Perovich, D. K., W. Meier, M. Tschudi, S. Gerland, and J. Richter-Menge. 2013. [The Arctic] Sea ice cover [in "State of the Climate in 2012"]. *Bulletin of the American Meteorological Society* 94(8): S126-S128.

Petersen, M.R., J.F. Piatt, and K.A. Trust. 1998. Foods of spectacled eiders *Somateria fischeri* in the Bering Sea, Alaska. *Wildfowl* 49: 124-128.

Petersen, M.R., W.W. Larned, and D.C. Douglas. 1999. At-sea distribution of spectacled eiders: A 120-year-old mystery resolved. *Auk* 116:1009-1020.

Petrich, C., P.J. Langhorne, and F.Z. Sun. 2006. Modelling the interrelationships between permeability, effective porosity and total porosity in sea ice. *Cold Regions Science and Technology* 44(2): 131-144.

Petrich, C., J. Karlsson, and H. Eicken. 2013. Porosity of growing sea ice and potential for oil entrainment. *Cold Regions Science and Technology* 87: 27-32.

Pew Charitable Trust. 2013. Arctic Standards: Recommendations on Oil Spill Prevention, Response, and Safety in the U.S. Arctic Ocean. Available, http://www.pewenvironment.org/uploadedFiles/PEG/Publications/Report/Arctic-Standards-Final.pdf, retrieved 24 November 2013.

Phillips, L.M., and A.N. Powell. 2009. Brood rearing ecology of king eiders on the North Slope of Alaska. *Wilson Journal of Ornithology* 121: 430-434.

Phillips, L.M., A.N. Powell, and E.A. Rexstad. 2006. Large-scale movements and habitat characteristics of king eiders throughout the nonbreeding period. *Condor* 108: 887-900.

Phillips, L.M., A.N. Powell, E.J. Taylor, and E.A. Rexstad. 2007. Use of the Beaufort Sea by king eiders breeding on the North Slope of Alaska. *Journal of Wildlife Management* 71: 1892-1898.

Pickart, R.S., T J. Weingartner, L.J. Pratt, S. Zimmermann, and D.J. Torres. 2005. Flow of winter-transformed Pacific water into the western Arctic. *Deep-Sea Research, Part II: Topical Studies in Oceanography* 52: 3175-3198.

Pickart, R.S., G.W.K. Moore, D.J. Torres, P.S. Fratantoni, R.A. Goldsmith, and J. Yang. 2009. Upwelling on the continental slope of the Alaskan Beaufort Sea: Storms, ice, and oceanographic response. *Journal of Geophysical Research: Oceans* 114(C1): C00A13.

Pickart, R.S., M.A. Spall, and J.T. Mathis. 2013. Dynamics of upwelling in the Alaskan Beaufort Sea and associated shelf-basin fluxes. *Deep Sea Research, Part 1: Oceanographic Research Papers* 1(76): 35-51.

Pisaric, M.F., J.R. Thienpont, S.V. Kokelj, H. Nesbitt, T.C. Lantz, S. Solomon, and J.P. Smol. 2011. Impacts of a recent storm surge on an Arctic delta ecosystem examined in the context of the last millennium. In *Proceedings of the National Academy of Sciences of the United States of America* 108(22): 8960-8965.

Pithan, F., and T. Mauritsen. 2014. Arctic amplification dominated by temperature feedbacks in contemporary climate models. *Nature Geoscience* 7: 181-184.

Polaris Applied Sciences. 2009. Guidelines and Strategies for Oil Spill Waste Management in Arctic Regions: Final Report. Available, http://arcticcouncil.org/eppr/wpcontent/uploads/2010/04/EPPRWasteManagement_FINAL Report_April2009.pdf.

Portnov, A., A.J. Smith, J. Mienert, G. Cherkashov, P. Rekant, P. Semenov, and B. Vanshtein. 2013. Offshore permafrost decay and massive seabed methane escape in water depths > 20 m at the South Kara Sea shelf. *Geophysical Research Letters* 40(15): 3962-3967.

Potter, R.E., J.T. Walden, and R.A. Haspel. 1981. Design and construction of sea ice roads in the Alaskan Beaufort Sea. Pp. 135-140 in *Offshore Technology Conference*. Houston, TX.

Potter, S. and I. Buist. 2010. In-situ burning in arctic and ice-covered waters: Tests of fire-resistant booms in low concentrations of drift ice. Pp. 743-754 in *Proceedings 33rd Arctic and Marine Oilspill Program Technical Seminar*, Environement Canada, Ottawa, Ontario.

Potter, S., I. Buist, K. Trudel, D. Dickins, and E. Owens. 2012. *Spill Response in the Arctic Offshore*, edited by D. Scholz. Prepared for the American Petroleum Institute and the Joint Industry Programme on Oil Spill Recovery in Ice. American Petroleum Institute, Washington, DC, 157 pp.

Powell, A.N, A.R. Taylor, and R.B. Lanctot. 2010. Pre-migratory and Physiology of Shorebirds Staging on Alaska's North Slope. Final Report, OCS Study MMS 2009-034, 194 pp.

Powell, S.M., M.J. Riddle, I. Snape, and J.S. Stark. 2005. Location and DGGE methodology can influence interpretation of field experimental studies on the response to hydrocarbons by Antarctic benthic microbial community. *Antarctic Science* 17: 353-360.

Power, G. 1997. A review of fish ecology in the Arctic North America. In *Fish Ecology in Arctic North America*, edited by J.B. Reynolds. American Fisheries Society, Bethesday, MD.

Preli, T., A.A. Allen, and D. Glenn. 2011. Development of high speed aerial ignition techniques for in situ burning. *Proceedings Arctic Technology Conference.*

Prince, R.C. 1993. Petroleum spill bioremediation in marine environments. *Critical Reviews in Microbiology* 19: 217-242.

Prince, R.C. 2005. The microbiology of marine oil spill bioremediation. Pp. 317-335 in *Petroleum Microbiology*, edited by B. Ollivier and M. Magot. ASM Press, Washington, DC.

Prince, R.C. 2010. Bioremediation of marine oil spills. Pp. 2617-2630 in *Handbook of Hydrocarbon and Lipid Microbiology*, edited by K.M. Timmins. Springer-Verlag, Berlin, Heidelberg.

Prince, R.C., and R.M. Atlas. 2005. Bioremediation of marine oil spills. pp. 269-292 in *Bioremediation: Applied Microbial Solutions for Real-World Environmental Cleanup*, edited by R.M. Atlas and J.C. Philp. *ASM Press*, Washington, DC.

Prince, R.C., and J.R. Clark. 2004. Bioremediation of marine oil spills. *Studies in Surface Science and Catalysis* 151: 495-512.

Prince, R.C., R.E. Bare, R.M. Garrett, M.J. Grossman, C.E. Haith, L.G. Keim, K. Lee, G.J. Holtom, P. Lambert, G.A. Sergy, E.H. Owen, and C.C. Guénette. 2003a. Bioremediation of stranded oil on an Arctic shoreline. *Spill Science and Technology Bulletin* 8: 303-312.

Prince, R.C., R.M. Garrett, R.E. Bare, M.J. Grossman, T. Townsend, J.M. Suflita, K. Lee, E.H. Owens, G.A. Sergy, J.F. Braddock, J.E. Lindstrom, and R.R. Lessard. 2003b. The roles of photooxidation and biodegradation in long-term weathering of crude and heavy fuel oils. *Spill Science and Technology Bulletin* 8: 145-156.

Prince, R.C., K.M. McFarlin, J.D. Butler, E.J. Febbo, F.C.Y. Wang, and T.J. Nedwed. 2013. The primary biodegradation of dispersed crude oil in the sea. *Chemosphere* 90: 521-526.

Prince William Sound Regional Citizens' Advisory Council. 2013. 2012-2013 Annual Report. Available, http://www.pwsr-cac.org/wp-content/uploads/filebase/about/annual_reports/currentannual/2012-2013%20PWSRCAC%20Annual%20 Report.pdf, retrieved 27 January 2014.

Pritchard, R., and D. Hanzlick. 1987. Chukchi Sea Ice Motion, 1981-82. POAC-87255-270. International Conference on Port and Ocean Engineering under Arctic Conditions, Fairbanks, AK.

Proshutinsky, A., R. Krishfield, M.-L. Timmermans, J. Toole, E. Carmackm, F. McLaughlin, and K. Shimada. 2009. Beaufort Gyre freshwater reservoir: State and variability from observations. *Journal of Geophysical Research: Oceans* (1978–2012), 114(C1): C00A10.

Provenzale, A. 1999. Transport by coherent barotropic vortices. *Annual Review of Fluid Mechanics* 31(1): 55-93.

Puestow, T., L. Parsons, I. Zakharov, N. Cater, P. Bobby, M. Fuglem, G. Parr, A. Jayasiri, S. Warren, and G. Warbanski. 2013. Oil Spill Detection and Mapping in Low Visibility and Ice: Surface Remote Sensing. Final Report by C-CORE for the Arctic Response Technology Joint Industry Programme.

Quakenbush, L.T., R. Suydam, R. Acker, M. Knoche, and J. Citta. 2009. Migration of King and Common Eiders past Point Barrow, Alaska, During Summer/Fall 2002 Through Spring 2004: Population Trends and Effects of Wind. Report to the Coastal Marine Institute, University of Alaska, Fairbanks, OCS Study MMS2009-036, 47 pp.

Quakenbush, L.T., J.J. Citta, J.C. George, R.J. Small, and M.P. Heide-Jørgensen. 2010. Fall and winter movements of bowhead whales (*Balaena mysticetus*) in the Chukchi Sea and within a potential petroleum development area. *Arctic* 63: 289-307.

Quakenbush, L.T., J.J. Citta, and J. Crawford. 2011. Biology of the Bearded Seal (*Erignathus barbatus*) in Alaska, 1961-2009. Final Report from Alaska Department of Fish and Game, Fairbanks, Alaska, to the National Marine Fisheries Service, 71 pp.

Quakenbush, L.T., R.J. Small, and J.J. Citta. 2013. Satellite Tracking of Bowhead Whales: Movements and Analysis from 2006-2012. U.S. Department of the Interior, Bureau of Ocean Energy Management, Alaska Outer Continental Shelf Region, Anchorage, AK. OCS Study BOEM 2013-01110, 60 pp + appendies.

Questel, J.M., C. Clarke, and R.R. Hopcroft. 2013. Seasonal and interannual variation in the planktonic communities of the northeastern Chukchi Sea during the summer and early fall. *Continental Shelf Research* 67: 23-41.

Reed, M., and O.M. Aamo. 1994. Real time oil spill forecasting during an experimental oil spill in the Arctic ice. *Spill Science and Technology Bulletin* 1(1): 69-77.

Rand, K.M., and E.A. Loggerwell. 2010. The first demersal trawl survey of benthic fish and invertebrates in the Beaufort Sea since the late 1970s. *Polar Biology* 34: 475-488.

Reed, M., D.P. French, T.A. Grigalunas, and J.J. Opaluch, 1989. Overview of a natural resource damage assessment model system for coastal and marine environments. *Oil and Chemical Pollution* 5: 85-97.

Reed, M., Ø. Johansen, P.J. Brandvik, P. Daling, A. Lewis, R. Fiocco, D. Mackay, and R. Prentki. 1999. Oil spill modeling towards the close of the 20th century: Overview of the state of the art. *Spill Science and Technology Bulletin* 5: 3-16.

Reed, M., Ø. Johansen, M. Moldestad, and F. Leirvik. 2009. Natural dispersion of heavy oil products and weathered crude oils. Presented at the 2009 Interspill Conference, Marseille, France.

Reed, M., I. Singsaas, P.S. Daling, L.-G. Faksnes, O.G. Brakstad, B.A. Hetland, and J.N. Hofatad. 2001. Modeling the water-accommodated fraction in OSCAR2000. *International Oil Spill Conference Proceedings* 2001(2): 1083-1091.

Regehr, E.V., S.C. Amstrup, and I. Stirling. 2006. Polar Bear Population Status in the Southern Beaufort Sea. U.S. Geological Survey Open File Report 2006-1337, 20 pp.

Regehr, E.V., C.M. Hunter, H. Caswell, S.C. Amstrup, and I. Stirling. 2007. Polar Bears in the Southern Beaufort Sea I: Survival and Breeding in Relation to Sea Ice Conditions, 2001-2006. U.S. Geological Survey, Alaska Science Center, Administrative Report, 45 pp.

Reimnitz, E., and E. Kempema. 1984. Pack ice interaction with Stamukhi Shoal Beaufort Sea, Alaska. Pp. 159-183 in *The Alaskan Beaufort Sea Ecosystems and Environments,* edited by P. Barnes, D. Schell, and E. Reimnitz. National Academy Press, Washington, DC.

Richard, P.R., A.R. Martin, and J.R. Orr. 2001. Summer and autumn movements of belugas of the eastern Beaufort Sea stock. *Arctic* 54: 223-236.

Ritchie, R.J., R.M. Burgess, and R.S. Suydam. 2000. Status and nesting distribution of lesser snow geese (*Chen caerulescens caerulescens*) and brant (*Branta bernicula nigricans*) on the western Arctic coastal plain, Alaska. *Canadian Field-Naturalist* 114: 395-404.

Ritchie, R.J., T. Obritschweitsch, R.M. Burgess, A.K. Prichard, and L.B. Attanas. 2013. Surveys for Nesting and Brood-Rearing Brant and Lesser Snow Geese, Barrow to Fish Creek Delta, Alaska, 2012. Unpublished report to the North Slope Borough, Department of Wildlife Management, Barrow, AK, 75 pp.

Roberts, K.H. 1990. Some characteristics of high-reliability organizations. *Organization Science.* 1: 160-177.

Röling, W.F.M., I.R.C. de Brito, R.P.J. Swannell, and I.M. Head. 2004. Response of archaeal communities in beach sediments to spilled oil and bioremediation. *Applied and Environmental Microbiology* 70: 2614-2620.

Romanovsky, V.E., S.L. Smith, H.H. Christiansen, N.I. Shiklomanov, D.S. Drozdov, N.G. Oberman, A.L. Kholodov, and S.S. Marchenko. 2011. Permafrost [in Arctic Report Card 2011]. NOAA. Available, http://www.arctic.noaa.gov/report11/permafrost.html, retrieved 25 October 2013.

Romanovsky, V.E., S.L. Smith, H.H. Christiansen, N.I. Shiklomanov, D.A. Streletskiy, D.S. Drozdov, N.G. Oberman, and S.S. Marchenko. 2012. Permafrost [in Arctic Report Card 2012]. NOAA. Available, http://www.arctic.noaa.gov/reportcard/ArcticReportCard_full_report.pdf, retrieved 25 October 2013.

Romanovsky, V.E., A.L. Kholodov, S.L. Smith, H.H. Christiansen, N.I. Shiklomanov, D.S. Drozdov, N.G. Oberman, and S.S. Marchenko. 2013. Permafrost [in "State of the Climate in 2012"]. *Bulletin of the American Meteorological Society* 94: S123-S124. Available, http://journals.ametsoc.org/doi/pdf/10.1175/2013BAMSStateoftheClimate.1, retrieved 25 October 2013.

Safine, D.E. 2013. Breeding Ecology of Steller's and Spectacled Eiders Nesting near Barrow, Alaska, 2012. Unpublished technical report, U.S. Fish and Wildlife Service, Fairbanks Fish and Wildlife Field Office, Fairbanks, AK, 64 pp.

SAIC (Science Applications International Corporation). 2011. Evaluation of the Use of Hindcast Model Data for OSRA in a Period of Rapidly Changing Conditions. Final Workshop Report. Report to BOEMRE. OSC Study BOEMRE 2011-032. Anchorage, AK.

Sambrotto, R.N., J.J. Goering, and C.P. McRoy. 1984. Large yearly production of phytoplankton in the western Bering Strait. *Science* 225(4667): 114(1): 1147-1150.

Schmutz, J.A. 2012. Monitoring Marine Birds of Concern in the Eastern Chukchi Nearshore Areas (Loons). Report to BOEM. OCS Study BOEM AK-07-04a, Anchorage, AK, 38 pp.

Schonberg, S.V., J.T. Clarke, and K.H. Dunton. 2014. Distribution, abundance, biomass and diversity of benthic infauna in the Northeast Chukchi Sea, Alaska: Relation to environmental variables and marine mammals. *Deep-Sea Research, Part II: Topical Studies in Oceanography* 102: 144-163.

Schulze, L.M., and R.S. Pickart. 2012. Seasonal variation of upwelling in the Alaskan Beaufort Sea: Impact of sea ice cover. *Journal of Geophysical Research: Oceans* 117(C1).

SEA Consulting Group. 2013. Dispersant Use in Ice-Affected Waters: Status of Regulation and Outreach Opportunities. Final Report 2.8. Joint Industry Programme.

Sergy, G.A., C.C. Guénette, E.H. Owens, R.C. Prince, and K. Lee. 2003. In-situ treatment of oiled sediment shorelines. *Spill Science and Technology Bulletin* 8: 237-244.

Serova, I. 1992. Behavior of oil in ice and water at low temperatures. In *Combatting Marine Oil Spills in Ice and Cold Climates*, HELLCOM Seminar, Helsinki, Finland.

Serreze, M.C., J.E. Walsh, F.S. Chapin III, T. Osterkamp, M. Dyurgerov, V. Romanovsky, and R.G. Barry. 2000. Observational evidence of recent change in the northern high-latitude environment. *Climatic Change* 46(1-2): 159-207.

Shakhova, N., I. Semiletov, A. Salyuk, V. Yusupov, D. Kosmach, and Ö. Gustafsson. 2010. Extensive methane venting to the atmosphere from sediments of the East Siberian Arctic Shelf. *Science* 327(5970): 1246-1250.

Sheffield, G., and J.M. Grebmeier. 2009. Pacific walrus (*Odobenus rosmarus divergens*): Differential prey digestion and diet. *Marine Mammal Science* 25: 761-777.

Sherr E.B., B.F. Sherr, and A.J. Hartz. 2009. Microzooplankton grazing impact in the western Arctic Ocean. *Deep-Sea Research, Part II: Topical Studies in Oceanography* 56(17): 1264-1273.

Singsaas, I., P.J. Brandvik, P.S. Daling, M. Reed, and A. Lewis. 1994. Fate and behavior of oil spilled in the presence of ice—A comparison of the results from recent laboratory, meso-scale flume and field tests. Pp. 355-370 in *Proceedings of the 17th Arctic and Marine Oilspill Program (AMOP) Technical Seminar*, Environment Canada, Ottawa, Ontario.

SL Ross Environmental Research Ltd. 1998. *Identification of Oils That Produce Non-Buoyant In-Situ Burning Residues and Methods for Their Recovery*. American Petroleum Institute and the Texas General Land Office, Washington, DC, 50 pp.

SL Ross Environmental Research Ltd. and MAR Inc. 2007. Corexit 9500 Dispersant Effectiveness Testing in Cold Water on Four Alaskan Crude Oils. U.S. Department of the Interior, Mineral Management Service.

SL Ross Environmental Research Ltd. and MAR Inc. 2008. U.S. Dispersant Effectiveness Testing on Viscous, U.S. Outer Continental Shelf Crude Oils: Phase II, Final Report. U.S. Department of the Interior, Minerals Management Service.

SL Ross Environmental Research Ltd., DF Dickins Associates Ltd., and Envision Planning Solutions Inc. 2010. Beaufort Sea Oil Spills State of Knowledge Review and Identification of Key Issues. Environmental Studies Research Funds Report No. 177. Calgary, AB, Canada, 126 pp.

Smith, L.C., and S.R. Stephenson. 2013. New trans-Arctic shipping routes navigable by midcentury. In *Proceedings of the National Academy of Sciences Early Edition*, 1-5. Available, www.pnas.org/cgi/doi/10.1073/pnas.1214212110, retrieved 4 October 2013.

Smith, M., N. Walker, C. Free, M. Kirchhoff, N. Warnock, A. Weinstein, T. Distler, and I. Stenhouse. 2012. Marine Important Bird Areas in Alaska: Identifying Globally Significant Sites Using Colony and At-Sea Survey Data. Audubon Alaska, Anchorage, 54 pp.

Smith, T.G. 1987. The ringed seal (*Phoca hispida*) of the Canadian western Arctic. *Canadian Bulletin of Fisheries and Aquatic Sciences* 216: 55.

Smith, V.A., D.W. Graham, and D.D. Cleland. 1998. Application of resource-ratio theory to hydrocarbon biodegradation. *Environmental Science & Technology* 32: 3386-3395.

Socolofsky, S.A., E.E. Adams, and C.R. Sherwood. 2011. Formation dynamics of subsurface hydrocarbon intrusions following the *Deepwater Horizon* blowout. *Geophysical Research Letters* 38(9): L09602.

Sørstrøm, S.E., P.J. Brandvik, I. Singsaas, S. Vefsnmo, H. Jensen, S.M. Løvås, M. Mathiesen, S. Løset, B.O. Johannessen, Ø. Johansen, P. Sveum, and C. Guénette. 1994. Experimental Oil Release in the Marginal Ice Zone. April 1993 (MIZ 93) Final Report. SINTEF. IKU Report 22.2120.00/02/94. 50 pp + 4 appendies.

Sørstrøm, S.E., P.J. Brandvik, I. Buist, P.S. Daling, D. Dickins, L.-G. Faksness, S. Potter, J.F. Rasmussen, and I. Singsaas. 2010. Joint Industry Program on Oil Spill Contingency for Arctic and Ice-Covered Waters. Summary Report-Oil in Ice JIP. SINTEF Report A14181.

Spaulding, M.L. 1988. A state-of-the-art review of oil spill trajectory and fate modelling. *Oil and Chemical Pollution* 4: 39-55.

Spaulding, M.L., V.S. Kolluru, E. Anderson, and E. Howlett. 1994. *Spill Science and Technology Bulletin* 1(1): 23-25.

Speckman, S.G., V.I. Chernook, D.M. Burn, M.S. Udevitz, A.A. Kochnev, C.V. Jay, A. Lisovsky, A.S. Fischbach, and R.B. Benter. 2011. Results and evaluation of a survey to estimate Pacific walrus population size, 2006. *Marine Mammal Science* 27: 514-553.

Spies, R.B. (ed.). 2011. An Independent Review of USGS Circular 1370: An Evaluation of the Science Needs to Inform Decisions on Outer Continental Shelf Energy Development in the Chukchi and Beaufort Seas, Alaska. Prepared for the Pew Environment Group and Ocean Conservancy.

Springer, A.M., and D.G. Roseneau. 1978. Ecological studies of colonial seabirds at Cape Thompson and Cape Lisburne, Alaska. National Oceanic and Atmospheric Administration and Bureau of Land Management. Pp. 839-960 in *Environmental Assessment of the Alaskan Continental Shelf, Annual Reports*, Vol. 2. Research Unit No. 460.

Springer, A.M., C.P. McRoy, and M.V. Flint. 1996. The Bering Sea Green Belt: Shelf-edge processes and ecosystem production. *Fisheries Oceanography* 5(3-4): 205-223.

St. Martin, J.W. 1987. Arctic Drifting Buoys Data, 1979-1985. Report No. CGR&DC07/87. U.S. Coast Guard R&D Center, Groton, CT.

Stafford, K.M., S.R. Okkonen, and J.T. Clarke. 2013. Correlation of a strong Alaska Coastal Current with the presence of beluga whales *Delphinapterus leucas* near Barrow, AK. *Marine Ecology Progress Series* 474: 287-297.

Stapleton, R.D., and G.S. Sayler. 2000. Changes in subsurface catabolic gene frequencies during natural attenuation of petroleum hydrocarbons. *Environmental Science & Technology* 34: 1991-1999.

Stegall, S.T., and J. Zhang. 2012. Wind field climatology, changes, and extremes in the Chukchi–Beaufort Seas and Alaska North Slope during 1979–2009. *Journal on Climate* 25: 8075-8089.

Stirling, I., M.C.S. Kingsley, and W. Calvert. 1982. The distribution and abundance of seals in the eastern Beaufort Sea, 1974-79. *Canadian Wildlife Service Occasional Paper*, No. 46, 27 pp.

Stoffyn-Egli, P., and K. Lee. 2002. Formation and characterization of oil-mineral aggregates. *Spill Science and Technology Bulletin* 8: 31-44.

Stralberg, D., D. Jongsomjit, C.A. Howell, M.A. Snyder, J.D. Alexander, J.A Weins, and T.L. Root. 2009. Re-shuffling of species with climate disruption: A no-analog future for California birds? *PLoS ONE* 4(9): e6825.

Stringer, W.J., S. Barrett, and L. Scheurs. 1980. Nearshore Ice Conditions and Hazards in the Beaufort, Chukchi, and Bering Seas. Geophysical Institute, University of Alaska, Fairbanks.

Stroeve, J.C., M.C. Serreze, M.M. Holland, J.E. Kay, J. Malanik, and A.P. Barrett. 2012. The Arctic's rapidly shrinking sea ice cover: A research synthesis. *Climatic Change* 110(3-4): 1005-1027.

Suydam, R.S. 2000. King eider (*Somateria spectabilis*). In *The Birds of North America*, edited by A. Poole and F. Gill. The Academy of Natural Sciences, Philadelphia, and The American Ornithologists' Union, Washington, DC.

Suydam, R.S. 2009. Age, growth, reproduction, and movements of beluga whales (*Delphinapterus leucas*) from the eastern Chukchi Sea. Ph.D. Dissertation. University of Washington, School of Aquatic and Fishery Sciences, Seattle, WA, 169 pp.

Suydam, R.S., L.F. Lowry, K.J. Frost, G.M. O'Corry-Crowe, and D. Pikok, Jr. 2001. Satellite tracking of eastern Chukchi Sea beluga whales in the Arctic Ocean. *Arctic* 54: 237-243.

Suydam, R.S., K.J. Frost, and L. Lowry. 2005. Distribution and Movements of Beluga Whales from the Eastern Chukchi Sea Stock During Summer and Early Autumn. Final Report OCS Study MMS 2005-035. University of Alaska, Coastal Marine Institute, Fairbanks, AK, 48 pp.

Sveum, P., and A. Ladousse. 1989. Biodegradation of oil in the Arctic: Enhancement by oil-soluble fertilizer application. *International Oil Spill Conference Proceedings,* 1989(1): 439-446.

Swannell, R.P.J., and F. Daniel. 1999. Effect of dispersants on oil biodegradation under simulated marine conditions. In *International Oil Spill Conference Proceedings,* 1999(1): 166-176.

Swannell, R.P., K. Lee, and M. McDonagh. 1996. Field evaluations of marine oil spill bioremediation. *Microbiological Reviews* 60: 342-365.

Szymoniak, N., and M. Devine. 2006. Wind Resource Assessment for Barrow, AK. Prepared by the Alaska Energy Authority, Anchorage. Available, http://www.akenergyauthority .org/PDF%20files/Wind%20Resource%20Assessment/Barrow_Wind-data%20report.pdf, retrieved 4 October 2013.

Teske, A., A. Durbin, K. Ziervogel, C. Cox, and C. Arnosti. 2011. Microbial community composition and function in permanently cold seawater from an Arctic fjord of Svalbard. *Applied and Environmental Microbiology* 77: 2008-2018.

Thienpont, J.R., D. Johnson, H. Nesbitt, S.V. Kokelj, M.F. Pisaric, and J.P. Smol. 2012. Arctic coastal freshwater ecosystem responses to a major saltwater intrusion: A landscape-scale palaeolimnological analysis. *The Holocene* 22(12): 1451-1460.

Thorsteinson, L.K. 1992. Arctic Fish Habitat Use Investigations: Nearshore Studies in the Alaskan Beaufort Sea, Summer 1990. OCS Study MMS 92-0011, Annual Report.

Timmermans, M.-L., and P. Winsor. 2013. Scales of horizontal density structure in the Chukchi Sea surface layer. *Continental Shelf Research* 52: 39-45.

Timmermans, M.-L., J. Toole, A. Proshutinsky, R. Krishfield, and A. Plueddemann. 2008. Eddies in the Canada Basin, Arctic Ocean, observed from Ice-Tethered Profilers. *Journal of Physical Oceanography* 38(1): 133-145.

Timmermans, M.-L., S. Cole, and J. Toole. 2012. Horizontal density structure and restratification of the Arctic Ocean surface layer. *Journal of Physical Oceanography* 42: 659-668.

Timmermans, M.-L., I. Ashik, Y. Cao, I. Frolov, R. Ingvaldsen, T. Kikuchi, R. Krishfield, H. Loeng, S. Nishino, R. Pickart, B. Rabe, I. Semiletov, U. Schauer, P. Schlosser, N. Shakhova, W.M. Smethie, V. Sokolov, M. Steele, J. Su, J. Toole, W. Williams, R. Woodgate, J. Zhao, W. Zhong, and S. Zimmermann. 2013. [The Arctic] Ocean Temperature and Salinity [in "State of the Climate in 2012"]. *Bulletin of the American Meteorological Society* 94(8): S128-S130.

Toole, J.M., M.L.Timmermans, D.K. Perovich, R.A. Krishfield, A. Proshutinsky, and J.A. Richter-Menge. 2010. Influences of the ocean surface mixed layer and thermohaline stratification on Arctic Sea ice in the central Canada Basin. *Journal of Geophysical Research: Oceans* (1978–2012): 115(C10).

Troy, D. 2010. Movements of Glaucous Gulls in Arctic Alaska 2009-2010. Unpublished report for the North Slope Borough, Department of Wildlife Management, 26 pp.

Tucker, W.B., W.F. Weeks, and M. Frank. 1979. Sea ice ridging over the Alaska continental shelf. *Journal of Geophysical Research* 84(C8): 4885-4897.

Una, G.V., and M.J.N. Garcia. 1983. Biodegradation of non-ionic dispersants in sea-water. *Journal of Applied Microbiology and Biotechnology* 18: 315-319.

UNEP (United Nations Environment Programme). 2012. Policy Implications of Warming Permafrost. UNEP: Nairobi, Kenya. Available, http://epic.awi.de/33086/1/permafrost.pdf, retrieved 25 October 2013.

UNOLS (United-National Oceanographic Laboratory System). 2013. AICC January 2013 U.S. Coast Guard Arctic Shield 2012 Native Outreach/Interactions. Available, http://www.unols.org/meetings/ 2013/201301aic/201301_aicap15.pdf, retrieved 17 August 2013.

USARC (U.S. Arctic Research Commission). 2012. Oil Spills in Arctic Waters: An Introduction and Inventory of Research Activities and USARC Recommendations. U.S. Arctic Research Commission and the U.S. Army Corps of Engineers: Cold Regions Research and Engineering Laboratory.

USARC. 2013. Report on the Goals and Objectives for Arctic Research 2013-2014. Prepared for the U.S. Arctic Research Program Plan.

USCG (U.S. Coast Guard). 2011. On Scene Coordinator Report, *Deepwater Horizon* Oil Spill. U.S. Department of Homeland Security, U.S. Coast Guard. Available, http://www.uscg.mil/foia/docs/dwh/fosc_dwh_report.pdf, retrieved 17 August 2013.

USCG. 2012. Oil Pollution Act Liability Limits in 2012: 2012 Report to Congress. U.S. Coast Guard.

USCG. 2013a. Arctic Shield, Oil Spill Response Joint Technology Demonstration Exercise. 7-20 September 2013. After Action Report, sponsored by USCG, BSEE, NOAA, and DHS Science and Technology Office of University Programs.

USCG. 2013b. Arctic Strategy. http://www.uscg.mil/seniorleadership/DOCS/CG _Arctic_Strategy.pdf, retrieved 5 July 2013.

USDOI (U.S. Department of the Interior). 2013. Review of Shell's 2012 Alaska Offshore Oil and Gas Exploration Program, March 8, 2013. Report to the Secretary of the Interior.

USFWS (U.S. Fish and Wildlife Service) Marine Mammals Management. 1999. Oil Spill Response Plan for Polar Bears in Alaska.

USFWS. 2003. Best Practices for Migratory Bird Care During Oil Spill Response, edited by Catherine Berg. U.S. Fish and Wildlife Service, Anchorage. Available, http://www.fws.gov/contaminants/fws_oscp_05/fwscontingencyappendices/D-BestPracticesMigBirds/BestPracticesmar04rev.pdf, retrieved 9 July 2014.

USFWS and USDOI. 2013. *Polar Bear News 2013-2014*. Available, http://alaska.fws.gov/fisheries/mmm/polarbear/pdf/2013%20PBNewsletter_Final.pdf, retrieved 6 September 2013.

USGAO (U.S. Government Accountability Office). 2012. Oil and Gas: Interior Has Strengthened Its Oversight of Subsea Well Containment, but Should Improve Its Documentation. *Report to Congressional Requesters.* GAO-12-244. Available, http://www.gao.gov/assets/590/ 588961.pdf, retrieved 10 September 2013.

Vandermeulen, J.H., and D.E. Buckley (eds.). 1985. The Kurdistan oil spill of March 16-17, 1979: Activities and observations of the Bedford Institute of Oceanography response team. Bedford Institute of Oceanography, Dartmouth, NS, Canada.

Varadaraj, R., M. L. Robbins, J. Bock, S. Pace, and D. MacDonald. 1995. Dispersion and biodegradation of oil spills on water. In *Proceedings of the 1995 International Oil Spill Conference, February 27-March 2, 1995, Long Beach, California*. pp. 101-106. American Petroleum Institute, Washington, DC.

Vaudrey, K., and D.F. Dickins. 1996. Oil Spill Tracking at the Northstar Development During Break-up and Early Summer Using Satellite-Tracking Buoys. Prepared for BP Exploration (Alaska) Inc., Anchorage, AK.

Venosa, A.D., and E.L. Holder. 2007. Biodegradability of dispersed crude oil at two different temperatures. *Marine Pollution Bulletin* 54: 545-553.

Venosa, A.D., and E.L. Holder. 2013. Determining the dispersibility of South Louisiana crude oil by eight oil dispersant products listed on the NCP Product Schedule. *Marine Pollution Bulletin* 66: 73-77.

Venosa, A.D., M.T. Suidan, B.A. Wrenn, K.L. Strohmeier, J.R. Haines, B.L. Eberhart, D. King, and E. Holder. 1996. Bioremediation of an experimental oil spill on the shoreline of Delaware Bay. *Environmental Science & Technology* 30: 1764-1775.

Venosa, A.D., K. Lee, M.T. Suidan, S. Garcia-Blanco, S. Cobanli, M. Moteleb, J.R. Haines, G. Tremblay, and M. Hazelwood. 2002. Bioremediation and biorestoration of a crude oil contaminated freshwater wetland on the St. Lawrence River. *Bioremediation Journal* 6: 261-281.

Vincent, W.F., and A. Quesada. 2012. Cyanobacteria in high latitude lakes, rivers and seas. P. 371 in *Ecology of Cyanobacteria II: Their Diversity in Space and Time*, edited by B.A. Whitton. Springer Science+Business Media B.V.

Voelker, R., and F. Seibold. 1990. Ice conditions along Alaskan marine transportation routes. Presented at ICETECH'90: Fourth International Conference on Ships and Marine Systems in Cold Regions, Calgary, AB, Canada.

von Biela, V.R., C.E. Zimmerman, and L.L. Moulton. 2011. Long-term increases in young-of-the-year-growth of Arctic cisco *Coregonus autumnalis* and environmental influences. *Journal of Fish Biology* 78: 39-56.

Wadhams, P. 1976. Sea ice topography in the Beaufort Sea and its effect on oil containment. *AIDJEX Bulletin 33*: 1-52.

Wadhams, P. 1981. Oil and ice in the Beaufort Sea--the physical effects of a hypothetical blowout. Pp. 299-308 in *Petroleum and the Marine Environment*. Graham and Trotman, London, UK.

Wadhams, P. 2000. *Ice in the Ocean*. Taylor and Francis, Philadelphia, 351 pp.

Wadhams, P. 2012a. Arctic ice cover, ice thickness and tipping points. *Ambio* 41(1): 23-33.

Wadhams, P. 2012b. New predictions of extreme keel depths and scour frequencies for the Beaufort Sea using ice thickness statistics. *Cold Regions Science and Technology* 76: 77-82.

Wadhams, P., and R.J. Horne. 1980. An analysis of the ice profiles obtained by submarine sonar in the Beaufort Sea. *Journal of Glaciology* 25(93): 401-424.

Wadhams, P., W.A. Squire, D.J. Goodman, A.M. Cowan and S. C. Moore. 1988. The attenuation rates of ocean waves in the marginal ice zone. *Journal of Geophysical Research* 93(C6): 6799-6818.

Wadhams, P., J.P. Wilkinson, and S.D. McPhail. 2006. A new view of the underside of Arctic sea ice. *Geophysical Research Letters* 33(4): L04501.

Wadhams, P., N.E. Hughes, and J. Rodrigues. 2011. Arctic sea ice characteristics in winter 2004 and 2007 from submarine sonar transects. *Journal of Geophysical Research* 116: C00E02.

Walker, N.D., C.T. Pilley, V.V. Raghunathan, E.J. D'Sa, R.R. Leben, N.G. Hoffmann, and R.E. Turner. 2011. Impacts of Loop Current frontal cyclonic eddies and wind forcing on the 2010 Gulf of Mexico oil spill. Pp. 103-116 in *Monitoring and Modeling the Deepwater Horizon Oil Spill: A Record-Breaking Enterprise*. Geopress, Belgium.

Wang, C., M. Quah, P.G. Noble, R. Shafer, K.A. Soofi, C. Alvord, and T. Brassfield. 2012. Use of jack-up drilling units in Arctic seas with potential ice incursions during open water season. Pp. 318-331 in *Society of Petroleum Engineers-Arctic Technology Conference 2012*, Volume 1. Curran Associates, Red Hook, NY.

Weeks, W.F., and W.D. Hibler III. 2010. *On Sea Ice*. University of Alaska Press, Fairbanks, 664 pp.

Weeks, W.F., W.B. Tucker III, M. Frank, and S. Fungcharoen. 1980. Characterization of Surface Roughness and Floe Geometry of Sea Ice over the Continental Shelves of the Beaufort and Chukchi Seas, from Sea Ice Processes and Model, edited by R. Pritchard. University of Washington Press, Seattle, WA.

Weingartner, T.J., D.J. Cavalieri, K. Aagaard, and Y. Sasaki. 1998. Circulation, dense water formation, and outflow on the northeast Chukchi shelf. *Journal of Geophysical Research* 103(C4): 7647-7661.

Weingartner, T.J., K. Aagaard, R. Woodgate, S. Danielson, Y. Sasaki, and D. Cavalieri. 2005. Circulation on the north central Chukchi Sea shelf. *Deep-Sea Research, Part II: Topical Studies in Oceanography* 52(24-26): 3150-3174.

Weingartner, T.J., S.L. Danielson, J.L. Kasper, and S.R. Okkonen. 2009. Circulation and Water Property Variations in the Nearshore Alaskan Beaufort Sea (1997-2007). Final Report. Institute of Marine Science, University of Alaska, Fairbanks.

Weingartner, T.J., E. Dobbins, S. Danielson, P. Winsor, R. Potter, and H. Statscewich. 2013a. Hydrographic variability over the northeastern Chukchi Sea shelf in summer-fall 2008–2010. *Continental Shelf Research 67: 5-22.*

Weingartner, T.J., P. Winsor, R. Potter, H. Statscewich, and E. Dobbins. 2013b. Application of High Frequency Radar to Potential Hydrocarbon Development Areas in the Northeast Chukchi Sea. OCS Study BOEM 2012-079, Final Report.

Whitehouse, G.A. 2013. Preliminary Mass-Balance Food Web Model of the Eastern Chukchi Sea. NOAA Technical Memorandum NMFS-AFSC-262.

Whiteman, G., C. Hope, and P. Wadhams. 2013. Climate science: Vast costs of Arctic change. *Nature* 499(7459): 401-403.

Whyte, L.G., A. Schultz, J.B. van Beilen, A.P. Luz, V. Pellizari, D. Labbe, and C.W. Greer. 2002. Prevalence of alkane monooxygenase genes in Arctic and Antarctic hydrocarbon-contaminated and pristine soils. *FEMS Microbiology Ecology* 41: 141-150.

Williams, W.J., E.C. Carmack, K. Shimada, H. Melling, K. Aagaard, R.W. Macdonald, and R. Grant Ingram. 2006. Joint effects of wind and ice motion in forcing upwelling in Mackenzie Trough, Beaufort Sea. *Continental Shelf Research* 26(19): 2352-2366.

Wilkinson, J., P. Wadhams, and N.E. Hughes. 2007. Modelling the spread of oil under fast sea ice using three-dimensional multibeam sonar data. *Geophysical Research Letters* 34: L22506.

Wilkinson, J., T. Maksym, and H. Singh. 2013. Capabilities for Detection of Oil Spills Under Sea Ice from Autonomous Underwater Vehicles. Polar Ocean Services and the Woods Hole Oceanographic Institution for the Arctic Response Technology Joint Industry Programme.

Wilkinson, J., T. Maksym, C. Bassett, A. Lavery, H. Singh, D. Chayes, and P. Jochmann. 2014. Oil Detection Under Sea Ice. Experiments Performed in the Experimental Basins at HSVA. Report to the European Commission from HSVA Ice Tank. Hamburg, Germany.

Woodgate, R.A., T. Weingartner, and R. Lindsay. 2010. The 2007 Bering Strait oceanic heat flux and anomalous Arctic sea-ice retreat. *Geophysical Research Letters* 37: 1-5.

Woshner, V.M, T.M. O'Hara, G.R. Bratton, and V.R. Beasley. 2001a. Concentrations and interactions of selected essential and non-essential elements in ringed seals and polar bears of Arctic Alaska. *Journal of Wildlife Diseases* 37: 711-721.

Woshner, V.M, T.M. O'Hara, G.R. Bratton, R.S. Suydam, and V.R. Beasley. 2001b. Concentrations and interactions of selected essential and non-essential elements in bowhead and beluga whales of Arctic Alaska. *Journal of Wildlife Diseases* 37: 693-710.

Woshner, V.M., T.M. O'Hara, J. Eurell, M.A. Wallig, G.R. Bratton, R.S. Suydam, and V.R. Beasley. 2002. Distribution of inorganic mercury in liver and kidney of beluga and bowhead whales through autometallographic development of light microscopic tissue sections. *Toxicologic Pathology* 30: 209-215.

Wrenn, B.A., J.R. Haines, A.D. Venosa, M. Kadkhodayan, and M.T. Suidan. 1994. Effects of nitrogen source on crude oil biodegradation. *Journal of Industrial Microbiology* 13: 279-286.

Wrenn, B.A., M.T. Suidan, K.L. Strohmeier, B.L. Eberhart, G.J. Wilson, and A.D. Venosa. 1997a. Nutrient transport during bioremediation of contaminated beaches: Evaluation with lithium as a conservative tracer. *Water Research* 31: 515-524.

Wrenn, B.A., M.T. Suidan, K.L. Strohmeier, B.L. Eberhart, G.J. Wilson, A.D. Venosa, J.R. Haines, and E. Holder. 1997b. Influence of tide and waves on washout of dissolved nutrients from the bioremediation zone of a coarse-sand beach: Application in oil-spill bioremediation. *Spill Science and Technology Bulletin* 4: 99-106.

Wrenn, B.A., K.L. Sarnecki, E.S. Kohar, K. Lee, and A.D. Venosa. 2006. Effects of nutrient source and supply on crude oil biodegradation, in continuous-flow beach microcosms. *Journal of Environmental Engineering* 132: 75-84.

WWF (World Wildlife Fund). 2010. Drilling for Oil in the Arctic: Too Soon, Too Risky. World Wildlife Fund, Washington, DC.

Yakimov, M.M., K.N. Timmis, and P.N. Golyshin. 2007. Obligate oil-degrading marine bacteria. *Current Opinion in Biotechnology* 18: 257-266.

Yapa, P.D., and D.P. Belaskas. 1993. Radial spreading of oil under and over broken ice: An experimental study. *Canadian Journal of Civil Engineering* 20(6): 910-922.

Yapa, P.D., and L.K. Dasanayaka. 2006. State-of-the-art review of modelling oil transport and spreading in ice covered waters. Pp. 893-909 in *Proceedings of the 29th Arctic and Marine Oilspill Program (AMOP) Technical Seminar*, Vol. 2. Environment Canada, Ottawa, Onario.

Yapa, P.D., and S.A. Weerasuriya. 1997. Spreading of oil spilled under floating broken ice. *Journal of Hydraulic Engineering* 12(8): 676-683.

Young, S. 2013. Meeting Arctic Missions. Department of Homeland Security, U.S. Coast Guard. Washington, DC, Available, http://coastguard.dodlive.mil/2013/07/meeting-arctic-missions/, retrieved 17 August 2013.

Zeller, D., S. Booth, E. Pakhomov, W. Swartz, and D. Pauly. 2011. Arctic fisheries catches in Russia, USA, and Canada: Baselines for neglected ecosystems. *Polar Biology* 34: 955-973.

Zhong, Y., and A. Bracco. 2013. Submesoscale impacts on horizontal and vertical transport in the Gulf of Mexico. *Journal of Geophysical Research: Oceans* 118(10): 5651-5668.

Zinjarde, S.S., and A.A. Pant. 2002. Hydrocarbon degraders from tropical marine environments. *Marine Pollution Bulletin* 44: 118-121.

ZoBell, C.E. 1973. Microbial degradation of oil: Present status, problems, and perspectives. Pp. 3-16 in *The Microbial Degradation of Oil Pollutants*, edited by D.G. Ahearn and S.P. Meyers. Center for Wetland Resources, Louisiana State University, Baton Rouge, LA.

A Committee and Staff Biographies

COMMITTEE

Martha Grabowski *(Chair)* is McDevitt Chair in Information Systems and Professor, Chair of the Business Administration Department at Le Moyne College, and Research Professor of Industrial & Systems Engineering at Rensselaer Polytechnic Institute. She is also a licensed former merchant officer and retired lieutenant commander in the U.S. Naval Reserve. Much of Dr. Grabowski's work centers on developing understanding of how large-scale systems of people, organizations, and technology behave, particularly those that are complex and geographically distributed. Her research projects focus on the impact of technology in safety-critical systems, risk analysis and risk mitigation in large-scale systems, and the role of human and organizational error in high-consequence settings. Currently, she is studying how social media can mobilize large masses of people in harm's way in very short time periods during extreme warning events, and modeling high-reliability virtual organizations for post-disaster cleanup, for global supply chains, and for financial cybersecurity systems. Dr. Grabowski is a recent past chair of the Marine Board and has served on numerous National Research Council (NRC) committees, including chairing the NRC committee on Naval Engineering in the 21st Century and vice-chairing the Committee on Review of the Tsunami Warning and Forecast System and Overview of the Nation's Tsunami Preparedness. Dr. Grabowski earned a B.S. from the U.S. Merchant Marine Academy and an MBA, an M.S. in engineering, and a Ph.D. in management and information systems from Rensselaer Polytechnic Institute.

Thomas S. Coolbaugh is a Distinguished Scientific Associate in the Oil Spill Response Technology Group at ExxonMobil Research and Engineering in Fairfax, Virginia. As the Oil Spill Response Technology Group lead, he provides technical guidance and training on the full suite of oil spill response strategies in support of global operations. In this position, his focus includes the use of dispersants, in situ burning, and remote sensing of oil spills. He has extensive experience in a variety of research settings as a research scientist and research leader and is an inventor/co-inventor on numerous U.S. and international patents covering a variety of compositions of matter and processes, including synthetic elastomers, dispersants for engine oils, and other specialty additives. Dr. Coolbaugh is chair of IPIECA's Oil Spill Working Group, and a member of the American Petroleum Industry (API) Spills Advisory Group and of the Marine Preservation Association Dispersant Advisory Committee. He is the author of various publications covering a range of subject matter including oil spill response technology. Dr. Coolbaugh received his B.A. in chemistry from Amherst College and his Ph.D. in chemistry from the California Institute of Technology. He has also received an M.S. in the management of technology from Polytechnic University (Now Polytechnic Institute of NYU).

David F. Dickins has over 40 years of project management experience focusing on environmental issues associated with offshore oil exploration and development, and marine transportation in Arctic waters. His company, DF Dickins Associates, founded in Canada's Northwest Territories in 1978, provides engineering research services for government and industry clients in the United States, Canada, Russia, Scandinavia, and Europe. Mr. Dickins has an established worldwide reputation and is regarded within the scientific and engineering community as an expert on Arctic sea ice and marine environments, oil spills in ice, remote sensing of ice, Arctic shipping route evaluations, and environmental impacts. He has worked on research issues related to cold water oil spill response and recovery for most of his career, including nine large-scale experimental field spills in ice and cold water covering the period from 1974 to 2009. Over the past 10 years, he led a series of joint projects funded by industry and the U.S. Minerals Management Service to test and develop new radar systems for oil-in-ice detection. In 2004 his firm was contracted to develop a set of Arctic oil spill research priorities for the U.S. Arctic Research Commission. He acted as project manager for remote sensing as part of the SINTEF Oil in Ice Joint Industry Project (JIP) in Norway from 2007 to 2010, and currently chairs the Field Research Technical Working Group as part of the ongoing Arctic Response Technology JIP. Mr. Dickins received a B.A.Sc. in mechanical engineering from the University of British Columbia, Vancouver in 1971 and is a registered professional engineer in the Province of British Columbia.

Richard K. Glenn is the Arctic Slope Regional Corporation's (ASRC's) Executive Vice President of Lands and Natural Resources and a member of their Board of Directors. The Arctic Slope Regional Corporation is an Alaska Native-owned regional corporation representing more than 10,000 Iñupiat Eskimos of Alaska's North Slope. The shareholders of ASRC own surface and subsurface title to nearly five million acres of Alaskan North Slope lands with oil, gas, coal, and mineral resources. Mr. Glenn has special expertise in resource development in an Arctic setting, and is well versed in on- and offshore Arctic geologic and sea-ice processes. From 1995 to 2001, he headed the North Slope Borough Department of Energy Management, where he supervised the energy programs for all of the North Slope Borough villages. He has twice been appointed by the President to the U.S. Arctic Research Commission and he is also the founder of the Barrow Arctic Science Consortium, a not-for-profit organization that encourages research and educational activities in the North Slope and facilitates the two-way transfer of information between scientists and the local communities. Mr. Glenn also has experience in his extended family's subsistence whaling crew, having served as its co-captain for almost 20 years. Mr. Glenn received a B.S. in geology from San Jose State University in 1985 and an M.S. in geology from the University of Alaska Fairbanks in 1991 and is a licensed professional geologist.

Kenneth Lee is the Director of the Wealth from Oceans National Research Flagship, part of Australia's Commonwealth Scientific and Industrial Research Organisation. He leads a national team of multidisciplinary scientists to promote research and development and the application of emerging technologies in ocean sciences to increase Australia's global competitiveness. He currently serves on the Australian Government National Plan for Maritime Environmental Emergencies Committee. Previously, he was Executive Director of the Centre for Offshore Oil, Gas and Energy Research, part of Fisheries and Oceans Canada. Dr. Lee's research and project management activities include studies to link organic and inorganic contaminants, marine noise, and alterations in hydrodynamic processes to effects on biota, including commercial fisheries species; chemical/microbiological studies on the biotransformation and biodegradation of contaminants; development of novel approaches to assess the impact of organic pollutants by the development and validation of toxicity assays based on advances in genomics, microbial ecology, and biochemical analysis; and coordination of multidisciplinary studies including the application of numerical models to predict the risk of industrial activities and contaminants on ecosystem health. Dr. Lee is one of the world's leading experts on the effects of dispersants and other spill response technologies. He recently served on the National Research Council Committee on the Effects of the Deepwater Horizon Mississippi Canyon-252 Oil Spill on Ecosystem Services in the Gulf of Mexico. Dr. Lee received a Ph.D. and M.Sc. in botany/environmental studies from the University of Toronto in 1982 and 1977, respectively, and a B.Sc. in biology from Dalhousie University in 1975.

William L. Majors has spent the past 10 years as the Planning & Development Manager for Alaska Clean Seas, a not-for-profit oil spill response cooperative located on the North Slope of Alaska. Their membership includes oil and pipeline companies that engage in or intend to undertake oil and gas exploration, development, production, or pipeline transport activities on the North Slope. Mr. Majors coordinates oil spill response research and development projects, training, and safety, including working with member companies on contingency plan development, oil spill response readiness, and oil spill response training and exercises and cooperating with regulatory and resource agencies on oil spill response readiness. Mr. Majors also spent almost 20 years in the U.S. Coast Guard, including 10 years' experience in marine safety and pollution response. He has over 25 years of oil spill response and management experience.

Mark D. Myers is the Vice Chancellor for Research at the University of Alaska Fairbanks (UAF), a position he has held since January 2011. Dr. Myers previously worked as the Alaska Gasline Inducement Act coordinator for the State of Alaska, and was the director of the U.S. Geological Survey from 2006 to 2009. His career as a geologist and policy maker spans more than three decades and includes work as a geologist for ARCO Alaska and the State of Alaska. He is an internationally recognized leader in energy and science policy and has wide-ranging industry, state, and federal government experience. From 2001 to 2005, he served as director of the state Division of Oil and Gas. Prior to his geology career, Dr. Myers served in the Air National Guard and Air Force Reserve as a pilot and intelligence officer. As Vice Chancellor for Research, Dr. Myers oversees administration of UAF's $123-million-per-year research enterprise and supervises the university's research institutes.

Dr. Myers received his B.S. and M.S. in geology from the University of Wisconsin-Madison in 1977 and 1981, respectively, and a Ph.D. in geology from UAF in 1994.

Brenda L. Norcross is a Professor of Fisheries Oceanography in the School of Fisheries and Ocean Sciences, UAF. Her research centers on fishes and their habitats, including human-induced effects on the environment. Since 2004, Norcross's research has focused on fishes in the Arctic, specifically the Chukchi and Beaufort Seas. Dr. Norcross headed the herring component of the multi-investigator Sound Ecosystem Assessment project, which investigated the environment of Prince William Sound following the *Exxon Valdez* oil spill. That research resulted in a synthetic knowledge of the juvenile life stage of herring. She has studied flatfishes in Alaskan waters and has modeled nursery habitats. Dr. Norcross was a member of two NRC committees, the Committee on Improving the Collection and Use of Fisheries Data and the Committee to Review the Gulf of Alaska Ecosystem Monitoring Program. Dr. Norcross earned her Ph.D. in marine science from the College of William and Mary in 1983.

Mark Reed is an expert in marine environmental modeling and impact assessments at SINTEF. SINTEF is the largest independent, nonprofit research organization in Scandinavia and has significant expertise in marine environmental technology, including fate and effects of spilled oil in cold regions. His scientific focus is on numerical model development for decision support applications, including coupled physical, biological, and chemical processes in the marine environment, with particular focus on fates and effects of pollutants (including oil, petroleum products, and chemicals). He has worked in quantitative environmental impact and natural resource damage assessment, and numerical modeling of animal migrations, biological transport, and fishery dynamics. Dr. Reed used his modeling expertise to help with the *Deepwater Horizon* spill in the Gulf of Mexico and has also worked on field programs for oil spill response in ice and cold water. He was operations manager at Applied Science Associates from 1982 to 1992, and group leader for Marine Environmental Modeling at SINTEF from 1992 through 2012. He now works as a private consultant supporting research and industrial organizations. Dr. Reed received a B.A. in philosophy from Antioch College in 1969, an M.S. in civil and urban engineering from the University of Pennsylvania in 1975, and a Ph.D. in ocean engineering from the University of Rhode Island in 1980.

Brian Salerno resigned from the committee in August 2013 to become the Director of the Bureau of Safety and Environmental Enforcement. Prior to this appointment, he served as the U.S. Liaison Officer for BIMCO, an international shipping association comprising a membership of a broad range of stakeholders with interests in the shipping industry. His previous U.S. Coast Guard positions included service as Deputy Commandant for Operations; and Assistant Commandant for Marine Safety, Security, and Stewardship. In these positions, VADM Salerno was responsible for national marine safety, security and environmental protection doctrine, policy, and regulations. He also ensured policy alignment with other federal partners and oversaw important work with federal advisory committees, industry-stakeholder partnerships, and international bodies. He received a master's degree in strategic studies from the U.S. Army War College in 2000, and also has a master's degree in management from the Johns Hopkins University.

Robert Suydam is a wildlife biologist whose research interests have focused on monitoring population trends and documenting natural history traits of bowhead whales, beluga whales, eiders, geese, caribou, and other Arctic species. He has held a position within the North Slope Borough's Department of Wildlife Management since 1990, and as part of this job conducts studies on wildlife species that are important for subsistence on the North Slope of Alaska. He is also responsible for reviewing documents related to oil and gas exploration, development, and production for projects onshore and offshore, with a focus on reviewing and evaluating impacts to bowhead whales and other marine mammals from industrial activities, particularly industrial sounds, in the Beaufort and Chukchi Seas off northern Alaska. He has written more than 50 peer-reviewed scientific publications and has authored numerous scientific reports. Dr. Suydam received a B.S. in environmental biology from California State University, Fresno, in 1986, an M.S. in biology from UAF in 1995, and a Ph.D. in aquatic and fishery sciences from the University of Washington in 2009.

James M. Tiedje is the University Distinguished Professor of Microbiology and Molecular Genetics, and of Plant, Soil and Microbial Sciences, and is Director of the Center for Microbial Ecology at Michigan State University. His research focuses on ecology, physiology, and genetics and genomics underlying important microbial processes in nature. Some of his relevant interests for this study include biodegradation of environmental pollutants and bioremediation, and microbial life in permafrost. His group has discovered several microbes that live by halorespiration on chlorinated solvents and is using genomics to better understand ecological functions, endemism, and niche adaptation. He has served as editor-in-chief of *Applied and Environmental Microbiology* and Editor of *Microbial and Molecular Biology Reviews*. He has over 450 refereed papers, including seven in *Science* and *Nature.* He shared the 1992 Finley Prize of UNESCO for research contributions in microbiology of international significance, is a fellow of the American Association for the Advancement of Science (AAAS), the American Academy Microbiology, and the Soil Science Society of America, and is a member of the U.S. National Academy of Sciences. He was president of the American Society for Microbiology in 2004-2005. He received his B.S. degree from Iowa State University and his M.S. and Ph.D. degrees from Cornell University.

Mary-Louise Timmermans is an Assistant Professor in the Department of Geology & Geophysics at Yale University. Her principal research focus is investigating the dynamics and variability of the Arctic Ocean to better understand how the ocean impacts Arctic sea ice and climate. Her approach is to apply fundamental theoretical models to geophysical observations, including measurements from an ice-based network of drifting automated ocean-profiling instruments, and hydrographic measurements from icebreaker surveys. Her research includes investigations of ocean mixing, eddies, waves, double-diffusive heat transport, and freshwater and heat content in the upper Arctic Ocean. She received a B.S. in physics at the University of Victoria, British Columbia, in 1994 and a Ph.D. in fluid mechanics from Cambridge University in 2000.

Peter Wadhams is Professor of Ocean Physics at Cambridge University. He is an oceanographer and glaciologist involved in polar oceanographic and sea ice research and concerned with climate change processes in polar regions. He leads the Polar Ocean Physics group, studying the effects of global warming on sea ice, icebergs, and the polar oceans. This involves work in the Arctic and Antarctic from nuclear submarines, autonomous underwater vehicles, icebreakers, aircraft, and drifting ice camps. He has more than 40 polar field expeditions. Dr. Wadhams' direct experience with oil spills in ice began during 1974-1976 with involvement, while with the Institute of Ocean Sciences in Victoria, British Columbia, in the Beaufort Sea Project, a large Canadian government project to study the effect of oil blowouts on ice in the Beaufort Sea. This included taking part in oil spills in the moving pack ice of the Beaufort Sea. He authored one of the Beaufort Sea Project reports. More recent relevant involvement includes the deployment of autonomous underwater vehicles under ice, and the European Union Interice project to carry out oil-ice studies using the Hamburg ice testing tank. Later, he was director of the Scott Polar Research Institute, Cambridge (1987-1992), before moving to his present position. He has been awarded the Polar Medal by the Queen of England and the Italgas Prize for Environmental Sciences. He is a member of the Scientific Committee of the European Environment Agency, Copenhagen, and a member of the Finnish Academy. Dr. Wadhams received his Ph.D. from the Scott Polar Research Institute in 1974.

STAFF

Deborah Glickson is a Senior Program Officer with the Ocean Studies Board at the National Research Council (NRC). She received an M.S. in geology from Vanderbilt University in 1999 and a Ph.D. in oceanography from the University of Washington in 2007. Her doctoral research focused on magmatic and tectonic contributions to mid-ocean ridge evolution and hydrothermal activity at the Endeavour Segment of the Juan de Fuca Ridge. In 2008, she participated in the Dean John A. Knauss Marine Policy Fellowship and worked on coastal and ocean policy and legislation in the U.S. Senate. Prior to her Ph.D. work, she was a research associate in physical oceanography at Woods Hole Oceanographic Institution. Since joining the NRC staff in 2008, she has worked on a number of ocean and earth science studies, including such topics as scientific ocean drilling, critical ocean science research needs and infrastructure, the academic research fleet, marine hydrokinetic energy, methane hydrates, and geoscience education.

Lauren Brown is an Associate Program Officer with the Polar Research Board and the Board on Atmospheric Sciences and Climate, where she has been involved in a number of studies such as *America's Climate Choices*, *Lessons and Legacies of International Polar Year 2007-2008*, and *Seasonal to Decadal Predictions of Arctic Sea Ice*. She holds an M.S. in Marine Studies with a focus on physical ocean science and engineering from the University of Delaware. She is especially interested in high-latitude environmental policy issues and the role of polar regions in global climate change.

Stacee Karras joined the NRC in September 2012 as a fellow in the Ocean Studies Board, and is currently a Research Associate. She received her B.A. in marine affairs and policy with concentrations in biology and political science from the University of Miami in 2007. The following year she received an M.A. in marine affairs and policy from the University of Miami's Rosenstiel School of

Marine and Atmospheric Science. Most recently, she earned her J.D. from the University of Virginia, School of Law.

Heather Chiarello joined the Ocean Studies Board in July 2008. She graduated magna cum laude from Central Michigan University in 2007 with a B.S. in political science with a concentration in public administration. She is pursuing a master's degree in sociology and public policy analysis at The Catholic University of America in Washington, DC. After five years with the Ocean Studies Board, Heather accepted a position as Senior Program Assistant with the Institute of Medicine in April 2013.

Payton Kulina joined the Ocean Studies Board as a Program Assistant in June 2013. He graduated from Dickinson College majoring in policy management, focusing on the Arctic National Wildlife Refuge and Rails-to-Trails projects. Prior to this position, Payton worked as a coordinator with BP Alternative Energy, also in Washington, DC.

B National Research Council Board Rosters

OCEAN STUDIES BOARD

MEMBERS

ROBERT A. DUCE, *Chair*, Texas A&M University, College Station
E. VIRGINIA ARMBRUST, University of Washington, Seattle
KEVIN R. ARRIGO, Stanford University, California
CLAUDIA BENETIZ-NELSON, University of South Carolina, Columbia
EDWARD A. BOYLE, Massachusetts Institute of Technology, Cambridge
RITA R. COLWELL, University of Maryland, College Park
SARAH W. COOKSEY, State of Delaware, Dover
CORTIS K. COOPER, Chevron Corporation, San Ramon, California
ROBERT HALLBERG, NOAA/GFDL and Princeton University, Princeton, New Jersey
DAVID HALPERN, Jet Propulsion Laboratory, Pasadena, California
SUSAN E. HUMPHRIS, Woods Hole Oceanographic Institution, Woods Hole, Massachusetts
BONNIE J. MCCAY, Rutgers University, New Brunswick, New Jersey
STEVEN A. MURAWSKI, University of South Florida, St. Petersburg
JOHN A. ORCUTT, Scripps Institution of Oceanography, La Jolla, California
H. TUBA ÖZKAN-HALLER, Oregon State University, Corvallis
STEVEN E. RAMBERG, Penn State Applied Research Lab, Washington, DC
MARTIN D. SMITH, Duke University, Durham, North Carolina
MARGARET SPRING, Monterey Bay Aquarium, Monterey, California
DON WALSH, International Maritime Incorporated, Myrtle Point, Oregon
DOUGLAS WARTZOK, Florida International University, Miami
LISA D. WHITE, University of California, Berkeley and San Francisco State University

Ex-Officio

MARY (MISSY) H. FEELEY, ExxonMobil Exploration Company, Houston, Texas

Staff

SUSAN ROBERTS, Board Director
CLAUDIA MENGELT, Senior Program Officer
DEBORAH GLICKSON, Senior Program Officer
STACEE KARRAS, Research Associate
PAMELA LEWIS, Administrative Coordinator
SHUBHA BANSKOTA, Financial Associate
PAYTON KULINA, Program Assistant

POLAR RESEARCH BOARD

Members

JAMES W. C. WHITE, *Chair*, University of Colorado, Boulder
WALEED ABDALATI, University of Colorado, Boulder
SRIDHAR ANANDAKRISHNAN, Pennsylvania State University
KATEY WALTER ANTHONY, University of Alaska Fairbanks
JULIE BRIGHAM-GRETTE, University of Massachusetts, Amherst
JOHN CASSANO, University of Colorado, Boulder
JENNIFER A. FRANCIS, Rutgers University
EILEEN E. HOFMANN, Old Dominion University
BERNICE M. JOSEPH, University of Alaska Fairbanks
ELLEN S. MOSLEY-THOMPSON, Ohio State University
GEORGE B. NEWTON, U.S. Arctic Research Commission
RAFE POMERANCE, Independent Consultant
CARYN REA, ConocoPhillips
GAIUS (GUS) R. SHAVER, Marine Biological Laboratory
ALLAN T. WEATHERWAX, Siena College

Ex-Officio

JACQUELINE M. GREBMEIER (U.S. Delegate to IASC), University of Maryland
TERRY WILSON (U.S. Delegate to SCAR), Ohio State University
DENEB KARENTZ (Alternate U.S. Delegate to SCAR), University of San Francisco

Staff

LAURIE GELLER, Senior Program Officer
LAUREN BROWN, Associate Program Officer

MARINE BOARD

Members

THOMAS M. LESCHINE, *Chair*, University of Washington
VICE ADMIRAL JAMES C. CARD, Ret., *Vice Chair*, Independent Consultant
STEVEN R. BARNUM, Hydrographic Consultation Services
MARY R. BROOKS, Dalhousie University
STEPHEN M. CARMEL, Maersk Line Ltd.
EDWARD N. COMSTOCK, Raytheon Company
ELMER P. DABENBERGER, III, Independent Consultant
REAR ADMIRAL THOMAS J. ECCLES, USJ-IMECO Holding Company
JEANNE M. GRASSO, Blank Rome, LLP
STEPHAN T. GRILLI, University of Rhode Island
CAPT. DOUGLAS J. GRUBBS, Crescent River Port Pilots Association
JOHN M. HOLMES, Independent Consultant
DONALD LIU, Marine Consultant
RICHARD S. MERCIER, Texas A&M University, College of Medicine
EDMOND J. MORAN, JR., Moran Towing Corporation
ALI MOSLEH, University of Maryland, College Park
CAPT. GEORGE B. NEWTON, JR., Independent Consultant
KARLENE H. ROBERTS, University of California, Berkeley
DAVID B. SANFORD, II, Manchester Maritime Associates, LLC
PETER K. VELEZ, Peter Velez Engineering, LLC

Staff

W. SCOTT BROTEMARKLE, Director
TIMOTHY DEVLIN, Senior Program Assistant

C Acronyms Used in the Report

ACADIS	Advanced Cooperative Arctic Data and Information Service
ADEC	Alaska Department of Environmental Conservation
AIS	Automatic Identification System
ANC	Alaska Native Corporation
AON	Arctic Observing Network
AOOS	Alaska Ocean Observing System
API	American Petroleum Institute
ASOS	Automated Surface Observing System
AUV	Autonomous Underwater Vehicle
BOEM	Bureau of Ocean Energy Management
BSEE	Bureau of Safety and Environmental Enforcement
CANUS	Canada-United States Joint Marine Pollution Contingency Plan
CANUSNORTH	Canada-United States Joint Marine Pollution Contingency Plan, Beaufort Sea Geographic Annex
COMIDA	The Chukchi Sea Offshore Monitoring in Drilling Area
CRREL	U.S. Army Cold Regions Research and Engineering Laboratory
DBO	Distributed Biological Observatory
DOD	U.S. Department of Defense
DWH	*Deepwater Horizon*
EPA	Environmental Protection Agency
ERMA	Environmental Response Management Application
ESI	Environmental Sensitivity Index

FEMA	Federal Emergency Management Agency
FLIR	Forward-Looking Infrared
FOSC	Federal On-Scene Coordinator
GIRG	Global Industry Response Group
GPR	Ground Penetrating Radar
GPS	Global Positioning System
HFR	High-Frequency Radar
HRVO	High-Reliability Virtual Organization
IARPC	Interagency Arctic Research Policy Committee
ICE-ARC	Ice, Climate, Economics—Arctic Research on Change
IR	Infrared
ISB	In Situ Burning
JIP	Joint Industry Program
LERPC	Local Emergency Response Planning and Coordination
LIDAR	Light Detection and Ranging
MARPOL	The International Convention for the Prevention of Pollution from Ships
MODIS	Moderate Resolution Imagine Spectroradiometer
NCP	National Contingency Plan
NEBA	Net Environmental Benefit Analysis
NIMS/ICS	National Incident Management System / Incident Command System
NMFS	National Marine Fisheries Service
NOAA	National Oceanic and Atmospheric Administration
NRC	National Research Council
NRDA	National Resource Damage Assessment
NRT	National Response Team
NSB	North Slope Borough
NSF	National Science Foundation
NSIDC	National Snow and Ice Data Center
NWAB	Northwest Arctic Borough
OCS	Outer Continental Shelf
OMA	Oil-Mineral Aggregate
OPA 90	The Oil Pollution Act of 1990

OPRC	International Convention on Oil Pollution Preparedness, Response, and Cooperation
OSLTF	Oil Spill Liability Trust Fund
OSRO	Oil Spill Removal Organization
PacMARS	Pacific Marine Arctic Regional Synthesis
PAH	Polycyclic Aromatic Hydrocarbon
PRAC	Primary Response Action Contractor
RP	Responsible Party
RRT	Regional Response Team
SAON	Sustaining Arctic Observing Networks
SAR	Synthetic Aperture Radar
SINTEF	Foundation for Scientific and Industrial Research (Norway)
SOAR	Synthesis of Arctic Research
SOSC	State On-Scene Coordinator
UAF	University of Alaska Fairbanks
UAV	Unmanned Aerial Vehicle
USCG	U.S. Coast Guard
USFWS	U.S. Fish and Wildlife Service
USGS	U.S. Geological Survey
VEC	Valued Ecosystem Component
VLOS	Very Large Oil Spill